Ancient Environments

Ancient Environments

Léo F. Laporte

Brown University

Prentice-Hall, Inc.,   Englewood Cliffs, New Jersey

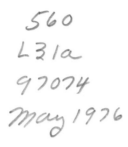
Design by Walter Behnke

Illustrations by Vincent Kotschar

PRENTICE-HALL INTERNATIONAL, INC., *London*

PRENTICE-HALL OF AUSTRALIA, PTY., LTD., *Sydney*

PRENTICE-HALL OF CANADA, LTD., *Toronto*

PRENTICE-HALL OF INDIA PVT. LTD., *New Delhi*

PRENTICE-HALL OF JAPAN, INC., *Tokyo*

Current printing (last digit):
10 9 8 7 6

FOUNDATIONS OF EARTH SCIENCE SERIES

A. Lee McAlester, Editor

C

For Mary

Foundations
of Earth Science Series

Elementary Earth Science textbooks have too long reflected mere traditions in teaching rather than the triumphs and uncertainties of present-day science. In geology, the time-honored textbook emphasis on geomorphic processes and descriptive stratigraphy, a pattern begun by James Dwight Dana over a century ago, is increasingly anachronistic in an age of shifting research frontiers and disappearing boundaries between long-established disciplines. At the same time, the extraordinary expansions in exploration of the oceans, atmosphere, and interplanetary space within the past decade have made obsolete the unnatural separation of the "solid Earth" science of geology from the "fluid Earth" sciences of oceanography, meteorology, and planetary astronomy, and have emphasized the need for authoritative introductory textbooks in these vigorous subjects.

Stemming from the conviction that beginning students deserve to share in the excitement of modern research, the *Foundations of Earth Science Series* has been planned to provide brief, readable, up-to-date introductions to all aspects of modern Earth science. Each volume has been written by an authority on the subject covered, thus insuring a first-hand treatment seldom found in introductory textbooks. Four of the volumes—*Structure of the Earth, Earth Materials, The Surface of the Earth,* and *Earth Resources*—cover topics traditionally taught in physical geology courses. Four more volumes—*Geologic Time, Ancient Environments, The History of the Earth's Crust,* and *The History of Life*—treat historical topics. The remaining volumes—*Oceans, Man and the Ocean, Atmospheres, Weather,* and *The Solar System*—deal with the "fluid Earth" sciences of oceanography and atmospheric and planetary sciences. Each volume, however, is complete in itself and can be combined with other volumes in any sequence, thus allowing the teacher great flexibility in course arrangement. In addition, these compact and inexpensive volumes can be used individually to supplement and enrich other introductory textbooks.

Acknowledgements

Grateful acknowledgement is here made to those persons who have been directly or indirectly influential in helping me write this book: Professor John Imbrie, Brown University, and Professor Norman D. Newell, Columbia University, who by their instruction, interest, and enthusiasm introduced me to the history and evolution of life; to Professor A. Lee McAlester, Yale University, who made helpful suggestions regarding the substance and expression of the ideas contained within this volume; to Mr. Ronald Nelson, Prentice-Hall, Inc., who gave considerable technical assistance; and to Dr. Alan Bé, Lamont Geological Laboratory, Professor Donald Hattin, Indiana University, and Dr. Max Pitcher, Continental Oil Company, who critically read and commented upon various parts of Chapter 6.

Illustration Credits

Figure 2–2 Fig. 139, p. 341, *Stratigraphic Principles and Practices* by J. Marvin Weller (Harper & Row, 1960). Figure 2–11 Thomas H. Clark and Colin W. Stearn, *The Geological Evolution of North America*—A Regional Approach to Historical Geology, Copyright © 1960, The Ronald Press Company, New York. Figure 2–14 From Bulletin of Geological Society of America, p. 1055 (1958). Figure 3–2 From Romer, *The Vertebrate Body,* 3rd edition, Philadelphia, W. B. Saunders Co., 1962. Figure 3–5 Adapted from E. Baldwin, *An Introduction to Biochemistry,* Cambridge University Press, 1948. In *Life: An Introduction to Biology,* 2nd edition, by George Gaylord Simpson and William S. Beck, © 1957, 1965 by Harcourt, Brace & World, Inc. and reproduced with their permission. Figures 3–7, 3–8b Courtesy American Museum of Natural History. Figures 3–9, 3–11 From *Life: An Introduction to Biology,* 2nd edition, by George Gaylord Simpson and William S. Beck, © 1957, 1965 by Harcourt, Brace & World, Inc. and reproduced with their permission. Figure 3–10 From Geological Society of America Memoir 67, v. 1, p. 37 (1957). Figure 3–13 (Below, left) Copyright 1963 by the American Association for the Advancement of Science. Figure 3–14 From Geological Society of America Memoir, 67, v. 1, p. 777 (1957). Figure 3–15 From Bulletin of Geological Society of America, v. 60, p. 375 (1949). Figure 5–1 From Geological Society of America Memoir, 67, v. 1, p. 408 (1957). Figure 5–2 From Geological Society of America Memoir 67, v. 1, p. 433 (1957). Figure 5–3 From Geological Society of America Memoir, v. 68, pp. 131–150 (1957). Figure 6–3 (left) From Bulletin of Geological Society of America 72 (1961). Figure 6–9 From R. C. Moore, *Introduction to Historical Geology,* 1958, 2nd edition. Used by permission of McGraw-Hill Book Company. Figure 6–10 From Geological Society of America Field Guidebook Annual Meetings, Kansas City, 1965, pp. 13, 19, 23. Figure 6–11 Reproduced from *Prehistoric Animals* by Josef Augusta and Zdenek Burian, Spring Books, London. Figures 6–19, 6–20 From M. Pitcher, 1964, Bulletin of Canadian Petroleum Geology v. 12, p. 642 (1964).

Contents

Ancient Environments

1

Geologic environments

Scientific explanation is often expressed in "if . . . , then . . ."
statements. That is, "if" certain necessary and sufficient conditions exist, "then" particular events will occur. For a simple
example: If water is cooled to 0°C at one atmosphere of pressure, then it will undergo a change in state from liquid to solid.
Similarly, the study of the Earth's environments is concerned
with establishing if–then relationships, by determining the
necessary and sufficient conditions required for diverse geologic phenomena.

There is a broad range of geologic environments that demands inquiry or definition. What are the pressures and temperatures deep in the Earth's crust? What assemblages of
silicate minerals will be at equilibrium under these temperatures and pressures? What are the states of stress and strain
in active mountain belts or in stable continental blocks? What
conditions of climate and habitat favored the invasion of land
by the first amphibians? The investigation of these and many
other geologic environments is being actively pursued by
Earth scientists today.

Although research in recent and ancient geologic environments and their associated geologic processes has proliferated lately, the foundations of environmental analysis are as old as the science of geology itself.

One of the earliest controversies in geology, almost two centuries ago, concerned the geologic environment responsible for the formation of basalt, a dark, fine-grained rock composed of various silicate minerals. Abraham G. Werner, a German mineralogist of the eighteenth century, insisted that basalt, like all other rocks, was deposited from a "universal ocean" that once covered the earth. Werner denied that rocks could form in any way except as chemical precipitates from this universal ocean. Two of Werner's students, D'Aubuisson de Voissins and Leopold von Buch, influenced by the work of a French geologist, Nicholas Desmarest, realized that some rocks, basalt in particular, had an igneous origin—that is, they had crystallized with the cooling of a molten rock mass. For many years the young science of geology was divided into two bitterly opposed camps: the "Neptunists" who argued that all rocks were water-laid, and the "Plutonists" who maintained that certain rocks owed their origin to the eruption of hot masses of molten rock to the surface from below the Earth's crust. The issue, of course, was the correct geologic environment for different kinds of rocks found at the Earth's surface.

What is new about the modern interest in analyzing geologic environments is the wide variety of investigative techniques that are available, many of which, such as those using the electron microscope and the mass spectrometer, were originally developed for other disciplines. In addition, contemporary geologists study existing geologic processes, whether natural or experimental, in order to understand ancient geologic processes and environments.

Ecology and Paleoecology

This volume will consider one part of the broad spectrum of geologic environments, that of recent and ancient sedimentary environments and their associated organic remains. The goal of these investigations is to understand the complex interrelationships between ancient organisms and their habitats. This area of Earth science is called *paleoecology*. It is related to the field of *ecology*, which is concerned with explaining the interaction of *living* animals and plants with their physical, chemical, and biological environment. Ecology is an established science with its own body of data, concepts, and principles.

Ecology, itself, is subdivided into two areas: *synecology* and *autecology*. Synecology attempts to relate the abundance and distribution of *whole faunas and floras* to particular environmental regimes. Autecology seeks to explain the interactions of a *specific group* of organisms within the fauna and flora with the local environmental conditions.

In a sense, paleoecology is merely ecology projected backward in time. In practice, however, paleoecologic studies are handicapped by the sparse preservation of ancient organisms. Many organisms that are preserved are extinct so that we do not know what their vital needs were during their lifetime. Additionally, various environmental factors in the ancient habitat, such as temperature, salinity, and humidity, are not directly recorded in the existing sedimentary rock. Consequently, paleoecologists have been forced to develop their own techniques and procedures for inferring from the enclosing rock matrix what the original environmental conditions may have been, as well as for estimating numbers of individuals and kinds of organisms in the ancient environment. Because of the inherent limitations of the subject and the relative newness of the field, paleoecology lacks, as yet, a coherent body of fundamental principles and concepts. In its current development the subject relies on empirically derived concepts, a multiplicity of investigative techniques, and a diversity of ideas and observations, all still essentially unintegrated.

Paleoecology and the Geological Sciences

Although it is intrinsically interesting to discover the kinds of environments in which various ancient fossil organisms flourished, paleoecology makes other important contributions to related geological fields of inquiry.

Paleoecology contributes most directly to *paleontology*, which is concerned with the history and evolution of life and therefore is a natural part of that science. For paleontologists seek more than merely a description of the various kinds of animals and plants that have lived in the past. They also wish to know why particular groups of organisms have evolved as they did and what environmental pressures the organisms were adapting to. To understand organic evolution, it is just as critical to know the habitats and habits of an organism as it is to know its shape and form.

Paleoecology provides information not only regarding the distribution of ancient lands and seas, but also about what sorts of terrestrial and marine environments these might have been. Such information is valuable to the fields of *sedimentation* and *stratigraphy*, which are concerned with the formation of sedimentary rocks and their ordering in time and arrangement in space.

These various interrelationships of organisms, environments, and geography are illustrated in Figure 1–1.

FIGURE 1–1 *The three factors of evolutionary stage, regional geographic distribution, and local environment determine the abundance and distribution of a local community of organisms. These factors apply to both a living community and a ancient fossil assemblage. The science of paleontology seeks to understand the role of each factor in explaining the history of life. The paleontologic specialties of paleosystematics, paleobiogeography, and paleoecology are concerned with each of these factors of evolution, biogeography, and local environment, respectively.*

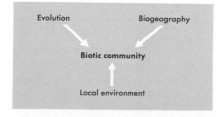

Classification of Environments

In defining and reconstructing ancient sedimentary environments, it is useful to consider the sorts of sedimentary environments present on the Earth today. The classification presented here is essentially based on physical criteria, although by implication chemical factors are involved as well. Moreover, since virtually all environments are populated by a variety of species adapted to the particular demands imposed by different environments, by extension, these environments will also presuppose certain biological factors. Thus environmental classification is by no means settled among ecologists as yet. There is still disagreement about the definition of environmental categories, particularly about the boundaries between environments. Such differences of opinion are natural because most environments merge into one another. Nevertheless, certain major environmental types are clearly distinctive. These categories are indicated below.

Marine Environments

The environments that begin at the sea's edge can be broadly subdivided into two major realms: the *pelagic*, which refers to the water mass itself, and the *benthic*, which refers to the substrate of sediments at the bottom. The pelagic realm can in turn be subdivided into the water that lies over the continental shelves (*neritic* environment) and the water mass that lies beyond the continental shelves in the deeper ocean basins (*oceanic* environment). The oceanic water mass can be still further differentiated into various other subenvironments according to depth of water.

The benthic realm has subdivisions, too, which correlate more or less with the pelagic subdivisions. Thus, the *sublittoral* environment includes the sea bottom on the continental shelves, and hence is overlain by the neritic pelagic environment. The sea floor beyond the continental shelves includes the *bathyal*, *abyssal* and *hadal* regions, which correspond more or less to the continental slope, the deep ocean floor, and the deep-sea trenches respectively.

That part of the sea floor that lies within the range of high and low tides is referred to as the *littoral* or *intertidal* environment. The narrow fringe of land that lies above the normal high water mark but is still within range of the sea's influence (salt spray, storm waves, or unusual high tides) is defined as the *supralittoral* environment. These various marine environments are illustrated in Fig. 1–2.

Terrestrial Environments

Terrestrial environments lying away from the sea's margin are considerably more diverse and individually more variable than are marine environments.

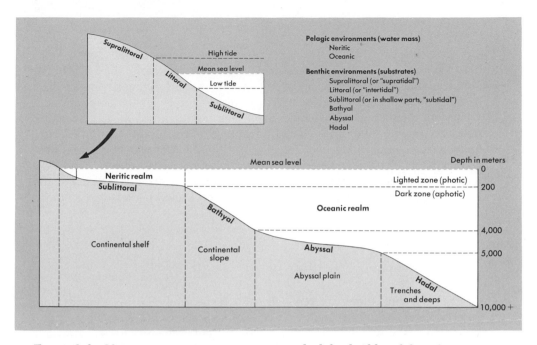

FIGURE 1–2 *Major marine environments as commonly defined. Although boundaries are not precise, note the general correspondence of sublittoral, bathyal, abyssal, and hadal environments with continental shelf, slope, abyssal plain, and trench. The penetration of sunlight decreases with water depth, being virtually absent in waters deeper than 200 meters, where it is always dark.*

For example, except for very shallow and isolated bodies of the sea, marine environments do not have the great fluctuations of temperature that occur on land.

On the land surface itself there are various aquatic environments, such as the *lacustrine* (lakes and ponds) and the *fluvial* (streams and rivers) as well as *swamps* and *marshes*. Two particularly important ecologic factors in these aquatic environments are the amount of water movement and the ratio of water surface area to water depth. Water movement, or current, influences the circulation of oxygen and nutrients. Where the currents are especially strong, the local inhabitants may have special adaptations that enable them to move upstream or that prevent their being swept downstream and eventually even out to sea. The ratio of water surface area to water depth influences the amount and distribution of dissolved oxygen within the water mass. In shallow streams and ponds there is usually sufficient oxygen distributed throughout the water to support a rich flora and fauna. Deep lakes, by contrast, may have inadequate quantities of dissolved oxygen because of a relatively small water surface area in contact with the atmosphere and a slow rate of circulation of the water mass as a whole. Consequently, many deep lakes will have bottom waters so low in oxygen that few, if any, organisms will be able to live there.

In nonaquatic terrestrial environments, physical conditions such as temperature, humidity, wind, and sunlight fluctuate considerably not only during the year but even daily. These environments are, therefore, intrinsically far more variable than terrestrial aquatic environments and still more variable than marine environments where ecological conditions are generally more constant.

Besides these temporal variations in physical conditions, there are also rather rapid geographic differences related to changes in topography, latitude, proximity to the oceans, and so on. The result is a number of widely different dryland habitats, including deserts, semi-arid plateaus, arctic tundra, and rain forests. These various habitats can be characterized in terms of their dominant physical conditions together with the associated organisms adapted to these conditions.

Plan of the Book

The following three chapters will deal, respectively, with sediments and environments, organisms and environments, and geochemical evidence of environments. We will then discuss how ancient environments are analyzed and how the resulting observations can be synthesized into meaningful generalizations about the conditions of life in the past. Finally, we will consider several ancient environments and illustrate how knowledge of them can be integrated into the broad field of historical geology. This volume, which emphasizes the environmental reconstruction of past events is complementary to other volumes in this series, particularly *The History of Life*, by A. Lee McAlester, which treats of the evolutionary development of ancient organisms, and *Geologic Time*, by Don L. Eicher, which considers the methods by which both ancient environments and fossil assemblages are dated within the sedimentary rock record.

2

Sediments and environments

Sediments are deposits of solid material made by a mobile medium—wind, ice, or water—at the surface of the Earth. These deposits are as varied as beach sands, lake muds, stream gravels, coral reefs, and desert dunes. The two main sources of this variety are the origin of the component grains and the environment in which these grains are laid down. Another important variable in the study of sediments is what happens physically and chemically to them *after* deposition—through the processes of compaction, cementation, and recrystallization—so that they become rocks. Therefore, we must consider three separate environmental influences responsible for the formation of sedimentary rocks. These are (1) the genesis of the sedimentary grains in the source area; (2) the transportation of these grains and deposition in their final resting place; and (3) the transformation of these loose grains into a compact, lithified, sedimentary rock.

Origin of the Sedimentary Grains

Sediments can initially be formed, either within, or outside of, the area where they are ultimately deposited. Sediments may, for example, be derived from the erosion of pre-existing rocks that lie at various distances from the place where they will eventually accumulate. Thus, the Mississippi River annually deposits more than 550 million tons of sediment which are derived from a very large region that covers all or part of 31 states, or 41 per cent of the total area of the contiguous United States. Coast lines, too, are continually eroded by the surf along the sea's margins. The resulting erosional products are deposited as local beaches and barrier bars or they are carried along the coast by long-shore currents and deposited a considerable distance from their place of origin.

Sediments also originate within the area of deposition. For example, various shelly invertebrates extract calcium carbonate from sea water in order to build their skeletons; on the death of the organisms their shelly remains are deposited with other accumulating sediments. In some hypersaline lakes and seas various salts such as sodium and calcium sulfates precipitate regularly because of high rates of evaporation. There are various places in the world—such as Pakistan, Russia, and Germany, as well as Kansas and Michigan in the United States—where ancient salt deposits are profitably mined for table salt, gypsum, and potassium. Besides these precipitation products, other sedimentary grains may be formed within a sedimentary basin.

It should be clear, therefore, that the composition of a particular sediment will usually reflect the composition of the rocks being eroded in the source area as well as the nature of the inorganic and organic precipitates that might also be forming in the area of sediment deposition.

The composition of sediments also depends on the *rates* of both weathering in the source area and deposition in the sedimentary basin. If rocks in the source area are deeply weathered, then their constituent minerals are chemically altered and mechanically disintegrated. If erosional rates are rapid, then the minerals are transported and buried before much alteration and disintegration can occur.

Consider the weathering of granite, which is composed of quartz, mica, and feldspar. If the rates of weathering and transport are relatively slow, the micas and feldspars will have ample time to break down into fine-grained clay minerals; the quartz grains may be rounded but they will not alter chemically because of the great stability of quartz. The resultant products of this granite therefore will be brought to the area of deposition as fine quartz sand mixed with finer-grained clays. The quartz sands may accumulate as nearshore or beach sands while the clays are carried farther offshore, where they eventually settle out of suspension as mud. If this same granite had been subjected to more rapid rates of weathering and transportation, then the resultant sedimentary deposits would have been quite different. The feldspars and micas would have been incompletely weathered and thus little altered, and would be deposited as fine sand mixed with the quartz grains, yielding a sediment of different composition and texture from that in the first case.

It should be noted in passing, too, that rates of erosion and transportation depend strongly on climate and topographic elevation of the source area. Table 2–1 illustrates some of the major sedimentary rocks classified according to type of weathering product, mode of deposition, and grain size.

Table 2–1

Origin and Classification of Major
Sedimentary Rock Types

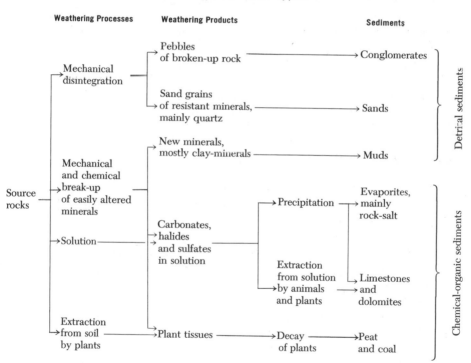

After H. H. Read and J. Watson, 1966.

Transportation and Deposition of Sediments

Water is the principal agent of sediment transport. When water falls as rain it is, at first, quickly absorbed by the soil. Soon, however, the upper layers of soil become saturated and the rain water begins to run off across the ground's surface. This surface runoff forms small rivulets that join to form brooks, streams, and eventually rivers.

During its overland journey the flow of water transports sedimentary materials in three different ways. First, the water takes into *solution* various substances such as calcium, iron, and carbonate ions. Second, fine-grained minerals and rock fragments are carried in *suspension* in the turbulent flow of the running water. Third, the water flow moves coarse-grained particles by *traction*, bouncing and rolling them along the stream or river bed.

The suspended and tractive sediments are eventually deposited in the delta at the river's mouth. Most of the dissolved load, however, is carried directly into the sea, where it may be either precipitated later (for example, by organisms as calcium carbonate) or where it may remain indefinitely in solution (for example, sodium chloride).

Ice, too, may be a significant agent of sediment transportation. For at various times in the past history of the Earth, thick masses of snow and ice covered large parts of the continents. As the snow collected and compacted, it turned to ice and under its own weight began to flow as a very viscous fluid, expanding outward from its place of initial accumulation. As these thick glacial masses inched inexorably forward they skimmed off soil and weathered rock layers. Some of this glacially eroded debris became frozen within the glacier only to be washed out, perhaps many miles distant from where it was first gathered up, when the glacier ice eventually melted. The hummocky topography in the northern latitudes of the northern hemisphere is the result of the deposition of sand and gravel by the Pleistocene glaciers, which were widespread in these areas.

Wind, the third agent, is a much less dense and viscous medium than either water or ice and therefore usually carries far less sedimentary material in suspension or traction, and virtually none in solution (although water vapor in the atmosphere may contain some dissolved salts). In areas where there is a poor cover of vegetation and where the climate is arid, there may be significant sediment transport by the wind, resulting in the formation of sand dunes. Windblown sand may also be a very effective erosional agent by abrading rock outcrops and desert pavements.

The transportation of sediments within the sea or along its margin is accomplished solely by moving sea water, although there may be occasional rafting of sediment out to sea by debris-laden icebergs. The movement of sedimentary grains within the oceans is basically the same as that in streams. Yet, although rivers and streams have fairly confined channels along which sedimentary particles are transported, the currents within the sea are often less well defined. For example, the Gulf Stream sweeps across the Blake Plateau, a particularly wide extension—more than 300 kilometers in some places—of the continental slope of the southeastern United States. Although the water depths over the Blake Plateau range from about 200 meters to more than 1000 meters, the surface of the plateau is only thinly veneered by recent marine sediments; rocks of Tertiary age crop out at or near the surface. Marine geologists have inferred from this that the broad surface of the Blake Plateau is being swept clean of any sedimentary material by the Gulf Stream whose axis of flow shifts periodically back and forth across the plateau.

Besides wind-induced and tidal currents, *turbidity currents* are also effective in the removal and transportation of marine sediments. Owing to the great topographic relief of the sea floors, intermittent movement of watery muds and sands by gravity flow occurs. Such turbid, sediment-laden currents, which are often triggered by earthquakes, can erode older, consolidated marine sediments. It has been suggested by several marine geologists that the submarine canyons that cut across the continental shelves and slopes have been carved out by such turbidity flows as material is moved from the continental margins down into the abyssal plains of the deep ocean basins.

There are often relatively long intervals during which sediments are not transported any significant distance in their journey from their source area to their final accumulation site. For the agents of sediment transportation vary in their capacity to carry materials and in their activity. For example, significant quantities of sediment may be transported only during the flood stage of a river. If so, between floods, just the dissolved fraction and the fine-grained, suspended fraction of a river's sediment load will move downstream. Sediments eroded in high, mountainous areas are deposited as alluvial fans within the adjacent valley floor where they can remain for long periods of time. But eventually, they, too, are eroded and retransported, gradually progressing toward the delta at their river's mouth. As for sediments that accumulate initially along the continental shelves, occasional turbidity currents will later transport some of this material farther out to sea in the deeper parts of the oceans. Therefore, although the net movement of sediment may be relatively slow by human standards, given the great eons of geologic time available for erosion, transportation, and deposition, the over-all impact of these geologic processes is enormous.

As indicated in Fig. 2–1, texture is a particularly useful characteristic in describing and interpreting a given sedimentary rock. Texture refers not only to the size of the component grains, but also to their shape and mutual arrangement within the enclosing matrix. Sedimentary textures often provide clues to the nature of the depositing medium. For example, the size and angularity of stream-laid deposits usually increase exponentially with an increase in the velocity of stream flow. Sediments that are coarse-grained, angular, and poorly sorted generally indicate rapid deposition by swift-moving water. On the other hand, sediments that are fine-grained, well-sorted, and laminated suggest deposition in quiet water, where individual grains settle slowly out of suspension. Combined with other evidence, textural analysis of a sedimentary rock can closely limit, if not actually define, the original depositional environment (Fig. 2–1).

It is important to emphasize, finally, that varying rates of weathering, trans-

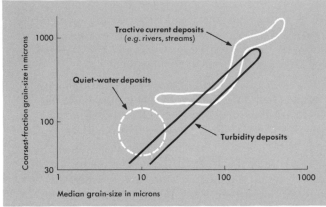

FIGURE 2-1 *Varying tex-tures of selected sedimentary deposits. Although there is some overlap of the fields, each tends to characterize a certain kind of depositional environment. The fields are defined by many samples from each environment. Co-ordinates for an individual sample are obtained from a size analysis of that sample. The "coarsest-fraction grain-size" is the diameter of grains in the coarsest 1 per cent (by weight), the median grain-size is the diameter of grains in the fiftieth percentile of the sample. Note that the co-ordinates are plotted loga-rithmically. (After R. Pas-sega, 1964.)*

portation, and deposition will influence the ulti-mate mineralogic and chemical composition as well as the texture of a sediment. Where weath-ering rates are rapid, the mineralogy and chemical composition of the resultant sediment are little altered and will be similar to that of the parent rocks in the source area. Where these rates are slow, however, the mineralogy and chemical com-position of the sediment will differ considerably from that of the source rocks. The texture, too, of the sediment will vary with amount of transportation of the grains to their area of accumulation. These various relationships are shown in Figs. 2–2 and 2–3.

Primary Structures

Besides differences in composition and texture, sediments often exhibit a variety of internal primary structures. Some of these primary structures are

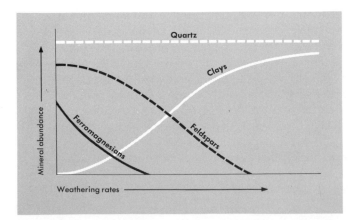

FIGURE 2-2 *Changes in the relative abundance of min-erals with increasing weath-ering rates. For this example granitic mineral constituents are shown: ferromagnesians (mica and some amphibole); feldspars; and quartz. Clays are chemical alteration prod-ucts of ferromagnesians and feldspars. (After J. M. Wel-ler, 1960.)*

directly related to the depositing medium and form as the sediments are laid down (ripple marks, for example). Others are merely incidentally associated with the local environment and form after the actual sediment deposition (mud-cracks). Some common primary structures of inorganic origin are described here, together with their environmental significance.

Cross-stratification

An internal layering of sedimentary grains that is inclined to the principal surface of deposition is called cross-bedding or cross-stratification (Fig. 2–4).

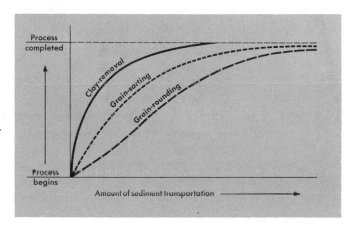

FIGURE 2–3 *Processes of clay removal, grain-sorting, and grain-rounding. The role of each process varies with the amount of sediment that has been transported. (After J. M. Weller, 1960.)*

Since the inclination of the cross-strata, which may be up to 30 degrees, points in the direction of local current movement, it is possible to determine from an analysis of cross-stratification within a sedimentary rock unit not only the general current direction but also the direction of the sediment's source. The geometry of cross-stratification is quite variable, but recent studies by geologists suggest that individual marine and nonmarine environments tend to have their own particular type of cross-stratification.

Ripple Marks

A surface of loose sediments may develop an undulating or rippled appearance as air or water currents move across it (Fig. 2–4). Where the current is moving uniformly from one direction to another, the ripple marks will be asymmetrical, with their steeper sides facing downstream (or downwind), while oscillating currents will form symmetrical ripples. Like cross-stratification, asymmetrical ripple marks can be used to infer former current directions.

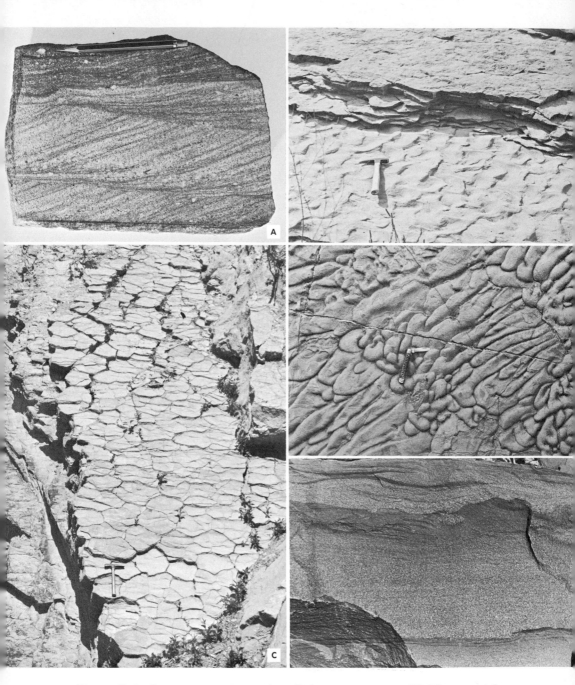

FIGURE 2–4 *Some common inorganic sedimentary structures. (A) Three sets of cross-stratification in a Mississippian sandstone where current moved from right to left. (B) Asymmetrical ripple marks in a Pennsylvanian sandstone where current moved from left to right. (C) Mud cracks in a Silurian limestone. (D) Sole markings on the bottom of an Ordovician turbidite bed where current flowed from lower left to upper right. (E) Three cycles of graded bedding in an Ordovician shale. (Courtesy F. J. Pettijohn and Paul Potter, Springer-Verlag New York, Inc.)*

Mud Cracks

Fine-grained, water-laid sediments that are later exposed to the air will usually shrink and crack as they dry out. These desiccation cracks form irregularly shaped polygons whose size is related to the thickness of the sedimentary layer being dehydrated (Fig. 2–4). Although some muds form shrinkage cracks under water ("syneresis cracks"), they do not have the connected, polygonal pattern of air-dried, mud-cracked sediments. Mud cracks, therefore, provide evidence of periodic exposure to air and, if combined with other relevant evidence, may indicate periods of temporary or long-continued terrestrial conditions.

Sole Markings

Deposition by turbidity currents may be preceded by local erosion of the sea floor as the debris-laden current moves across it. The scour marks that are formed by the turbidity current have a variety of characteristic forms, some of which indicate the direction from which the current came. The scour marks are actually preserved as casts formed by the sediment that is deposited by the turbidity current after its eroding front has passed by. The casts of these scour marks thus appear on the sole, or bottom, of the turbidity-laid sedimentary layer, hence the term "sole markings" (Fig. 2 4).

Graded Bedding

The sediment deposited by a turbidity current is usually laid down in such a way that the coarse grains are dropped first, followed by the settling out of the finer grains. This change in texture occurs because as the turbidity current's velocity decreases, the coarsest pebbles and sand grains will be deposited initially. This coarse layer will then be slowly buried by finer-grained sand, silt, and clay as they settle out of suspension from the overlying turbid water. This regular variation in grain size from the base of a sedimentation unit to the top is called graded bedding (Fig. 2–4). Although graded bedding can also be found in other sedimentary environments, it is quite typical of rock sequences laid by turbidity currents and combined with other criteria, such as sole markings, can be useful in defining turbidity-current environments.

Diagenesis

The appearance and properties of a sedimentary rock are largely due to what happens to a sediment after its initial deposition. Those processes responsible

Sediments and environments

for making a loose, wet sediment into a compact, dense rock are included within the general term diagenesis. Just as the original depositional environment influences the character of a given sedimentary rock, so too does the post-depositional environment influence its diagenesis.

After a sediment is laid down, it is usually buried by successive deposits as the area of sedimentation continues to subside. As burial continues, the weight of the overlying, accumulating sediment causes the materials below to undergo *compaction* and *consolidation*. Individual sedimentary grains are pushed and crowded together, thus reducing the initial pore space that existed between the grains. In the case of water-laid sediments, the interstitial fluids are slowly squeezed out as the porosity of the sediment diminishes—sometimes resulting in a volume decrease of 50 per cent or more. The interstitial fluid is originally incorporated from the transporting medium (sea water or fresh water) during sedimentation, and is identical with it in terms of chemical composition, acidity, and oxidizing potential. With time, however, the interstitial water may radically change its chemical character. Much of this change is due to the solution of unstable minerals and the decomposition of included organic matter by micro-organisms.

As the coarser-grained sediments are compacted, the pore water moves upward through the sediments and may dissolve or precipitate mineral matter along the way. Thus, compaction of sediments with subsequent squeezing out of the interstitial fluids initiates the next step in the diagenetic process, namely, *grain cementation*. Individual sedimentary grains are welded together by mineral matter—usually silica or calcium carbonate—deposited by the interstitial fluids. Cementation may continue, especially if the sediments are later uplifted and located within the zone of fresh water percolation and saturation where large quantities of fluids may freely circulate.

Muddy sediments, on the other hand, after compaction, undergo *recrystallization,* which is likely to be more prevalent in these finer-grained sediments than grain-to-grain cementation. Although the kinetics of this process are not fully understood, it appears that as muddy sediments are consolidated, individual mineral grains are brought close together and, through a reorientation of their atoms and molecules, recrystallize into a dense and almost impermeable framework. This spontaneous recrystallization may be triggered by chemical equilibrium changes brought about with increasing pressures and temperatures as the sediments are buried deep within the upper part of the Earth's crust.

Organic Influences on Sediments

Organisms can exert strong influences on sediments by contributing sedimentary grains in the form of skeletal debris, by producing primary structures, and by not only altering the original composition and textures of sediments but also modifying patterns of sedimentation through their activities. In fact, it has only been in recent years that geologists have had a full awareness of the myriad ways and great potential of organisms for influencing sedimentation.

Organic Contributions to Sediments

Organisms contribute directly to sediments by producing in the case of animals, a variety of internal and external skeletal materials such as bone, teeth, shells, and in the case of plants, woody tissue, all of which become sedimentary grains after the death of the organisms. Of these, however, only the calcareous- and siliceous-secreting protistans (such as diatoms, radiolarians, and foraminifers), and the calcareous algae and invertebrates (including corals, brachiopods, bryozoans, molluscs, and echinoderms) have any real quantitative significance in the sedimentary record. Thus, many rock formations of the last half billion years or so are anywhere from one to more than 75 per cent biological in origin, and thereby providing the historical record of life on earth.

Skeletal materials are weathered, transported, and deposited much the same as inorganically formed rocks and minerals are. Most skeletal structures are secreted in an organic matrix that decomposes after the death of its owner. The organic matrix is attacked by micro-organisms and is oxidized by oxygen in the atmosphere or in water. Consequently, the individual crystalline units, which compose the skeleton and are imbedded in this matrix, are liberated and shed into the sediment. For example, a large clam shell which begins as a large, cobble-sized sedimentary grain will, as the binding matrix of the shell is removed, eventually break down into many thousands of tiny calcite prisms and aragonite needles just a few microns long. Depending on the degree of skeletal decomposition, shelly sediments will thus have textures that reflect the original internal architecture of the shell materials contributed by the local organisms (Fig. 2–5).

After deposition, the organically produced crystalline remains are susceptible to solution, particularly some of the more unstable mineral types, such as aragonite and opal, and provide a reservoir of calcium carbonate and silica for subsequent sediment cementation. Differing solubility of various kinds of skeletal debris can further bias any estimate about the original composition of the local flora and fauna based on fossil remains within a sedimentary rock. (The initial bias introduced, of course, is that of the nonpreservation of the many soft-bodied organisms that secrete no mineral material whatsoever.)

In short, then, through their secretion of various kinds of skeletal materials, organisms can contribute directly and significantly to the ultimate composition and texture of a sediment. And because organisms in turn are limited in their distribution and abundance by the local environment (as will be discussed in the next chapter), the influence that the depositional environment exerts on the character of the accumulating sediments is demonstrated once again.

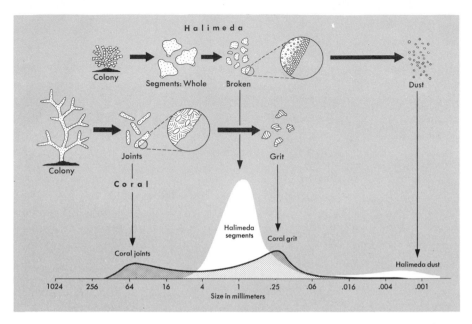

FIGURE 2–5 *How a green calcareous alga (Halimeda) and a stony coral (Acropora) contribute skeletal debris to sediments. Note the wide variation in size from a few centimeters to several microns, depending on the degree of disarticulation and disintegration of the original skeleton. These two organisms thus can produce calcareous sediments of various textures from coral gravel to lime mud. (After R. Folk and R. Robles, 1964.)*

Organic Alterations of Sediments

Organisms contribute in other, less direct ways to sediments. Many marine organisms, especially worms, arthropods, and molluscs, burrow into sediments for shelter and food (Fig. 2–6). By so doing they disrupt bedding, obliterate

FIGURE 2–6 *(A) Artificially laminated sediment with alternating layers of fine (dark) and coarse (light) sand in an aquarium. (B) Five marine worms were placed within the aquarium and after one month the four-and-one-half-inch bed of sediment has been stirred up and the bedding has been obliterated. (Courtesy R. N. Ginsburg, S.E.P.M. Special Publication No. 5.) (C) Natural view of a marine echinoid burrowing through a rippled, calcareous sand. (Courtesy Norman D. Newell, American Museum of Natural History, S.E.P.M. Special Publication No. 5.)*

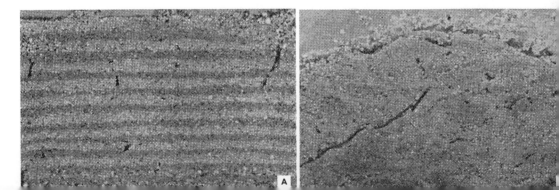

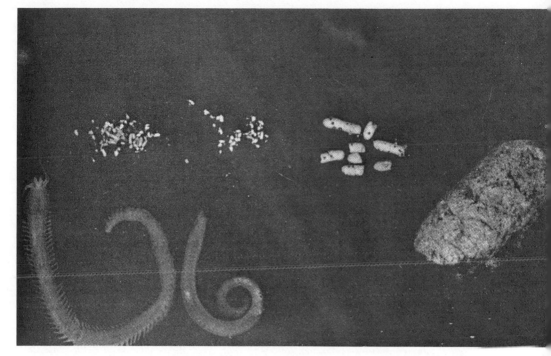

FIGURE 2-7 *Representative fecal pellets formed by various marine invertebrates. The two examples in upper left are formed by worms, the pellets in the center are from an intertidal mollusc (chiton), while the large pellet at the right is from a crustacean. (Courtesy R. N. Ginsburg)*

primary structures, and increase sea water circulation within the sediments. Organisms that feed on sediment for its included organic matter may aggregate the sediment to form pellets as it passes through the digestive tract (Fig. 2–7). In environments where sediment is reworked by organisms faster than it accumulates—because of either slow sedimentation rates or large populations of organisms—the resulting sediments may be so extensively modified that they contain few traces of the original inorganic environment.

Besides modifying the character of sediments already deposited, organisms are also capable of altering patterns of sedimentation. For instance, thin mats composed of the intertwined filaments of various kinds of blue-green and green algae can exert local, small-scale influence. These algal mats, which oc-

cur in very shallow water, both fresh and marine as well as on moist land surfaces, form tough, leathery coverings on the sediment surface and thus inhibit or prevent the transport of grains along the surface. Hence, the mats act to keep sediments stable and free from erosion after their deposition.

Algal mats are also capable of trapping and binding sediment grains on their gelatinous surface. As the mats are covered by newly deposited sediments, the algal filaments grow upward and form yet another mat. Successive periods of matting and deposition can result in a well-laminated sediment (Fig. 2–8). Furthermore, the geometry of these laminations is related to the frequency and strength of local water movement.

Some of the oldest sedimentary rocks known contain laminated structures that are virtually identical with recent algal laminated sediments. These rock structures, termed *"stromatolites,"* are believed to be algal in origin, the layering or laminations being related to the successive development of algal mats, although the algal tissue itself has not been preserved (Fig. 2–8). The

Figure 2–8 *(A) Vertical sketch of algal laminated sediment; algal mats localized as small mounds which grow upward as successive layers of sediment are trapped by successive generations of mat formation. (B) A number of small algal mounds on floor of saline lake in western Australia. During part of year the lake floor is covered with water and the algal mats are rejuvenated; hammer gives scale. (Courtesy Brian W. Logan.) (C) Large, single algal mound from intertidal zone of Shark Bay, Australia; coarse shell debris surrounds mound. (Courtesy Brian W. Logan.) (D) Cross section of a fossil algal mound (or stromatolite) from Precambrian rocks, about 1 billion years old, of Montana. (Courtesy Richard Rezak, U.S. Geological Survey.)*

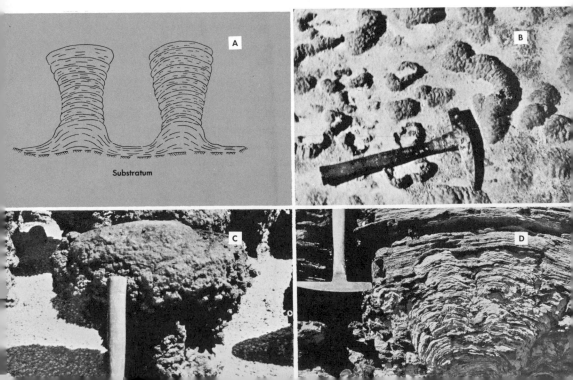

Substratum

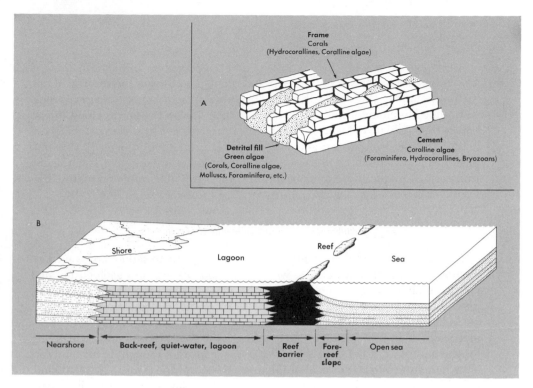

FIGURE 2–9   *(A) Schematic drawing of the role of various calcareous organisms in reef construction. (After R. N. Ginsburg and H. A. Lowenstam, 1958.) (B) Ancient reefs often had different kinds of organisms but they performed much the same structural function. As a reef develops, local patterns of sedimentation are altered; major subenvironments within a reef complex are the reef barrier and the environments in front of (more turbulent) and behind (less turbulent) the barrier.*

presence of stromatolites in many Precambrian rocks, some of which are dated at 2 billion years or more, indicates the great antiquity of life—even if only "primitive" life. Stromatolites are also useful environmental indicators because rocks containing them were presumably deposited in very shallow water, if we can assume that algal mats occurred in the same environments in the past as they do today.

Other, and geologically more dramatic, examples of the organic modification of local patterns of sedimentation are provided by reefs. Organic reefs developed in those parts of ancient seas that were clear, shallow, warm, well lit, and agitated, so that a profusion of shelly marine invertebrates flourished. Some of these organisms—such as stony corals, calcareous sponges, and certain rooted brachiopods—acted as *frame-builders* of these reefs. Other organisms—such as calcareous algae and hydrocoralline coelenterates—because of their encrusting mode of growth acted as the *cement*, binding the framework of the reef together into a rigid, wave-resistant structure. Other marine invertebrates—such as molluscs and echinoderms—provided the *detrital fill* for the growing reef mass (Fig. 2–9).

Reefs can perpetuate the favorable conditions for their initial formation by growing as fast as the sea floor subsides, thereby maintaining the survival of the reef community. With time the reefs become large masses of calcareous rock surrounded by nonreef sediments. In some cases the reefs will form long, linear bodies that build up to mean sea level. This type will have a quieter water lagoon behind the reef-wall proper, and a turbulent fore-reef environment. Thus, what began as an essentially uniform sedimentary marine environment is gradually transformed by reef development into several highly differentiated marine environments: *back-reef lagoon, reef barrier,* and *fore-reef slope* (Fig. 2–9).

Significance of Organic Influence on Sediments

Having established that organisms contribute grains to sediments as well as disrupt their internal structures, we may reasonably ask how geologically significant such activity is. That is, given the rate at which these various organic processes occur, how great will the effects be in the geological record?

Sedimentary rocks constitute about 75 per cent areally and 5 per cent volumetrically of rocks within the Earth's crust. Of these, limestones—which are for the most part, if not entirely, biologic in origin—are thought to account for anywhere from 5 to 20 per cent of the total. Hence, of the estimated 400 million cubic kilometers of sedimentary rock, 20 to 80 million are the result of calcareous-secreting and -precipitating organisms. Although this is a startling figure in absolute terms, given the great lapse of geologic time, slow depositional rates can easily account for such a great mass of biogenic calcareous sediments. For example, the present-day Bahama Islands in the Caribbean are underlain by a section of shallow-water, shelly sediments that are at least 4,500 meters thick and that go back to the late Cretaceous Period. During the last 100 million years, therefore, some 4,500 meters of limestone have been deposited, giving an average accumulation rate in this area of only about 4.5 centimeters per 1,000 years. Such a rate surely must be minimal, however, because sedimentation has not been continuous owing to times of lowered sea level—as during the Pleistocene glaciations—when the Bahamas stood above water and were subjected to erosion. Present sedimentation rates as measured since the last inundation of the shallow Bahama platform indicate that the true rate may be as much as eight to nine times as great.

Estimates can also be made for accumulation rates of the minute calcareous shells of floating protistans that rain down on the ocean floors. For example, a core raised from deep-water sediments from an area south of the island of Hispaniola measures more than 900 centimeters, representing sediment accumulation during about 400,000 years. An average sedimentation rate for this Caribbean core is thus calculated at almost 2.5 centimeters per 1,000 years.

How do these shallow and deep-water rates of carbonate sedimentation compare with areas where no carbonate is deposited? In the Gulf Coast sedimentary trough, or geosyncline, sedimentation rates are estimated at 20 centimeters per

1,000 years as averaged over the Cenozoic Era. Once again, this is probably too low a figure because this area has, at times, been subject to erosion and so has not continuously received sediments. Depositional rates for red clays in the deep ocean basins are quite low, ranging from one-twentieth to one-half centimeter per 1,000 years. As shown in Table 2-2, sediments of organic origin are indeed being deposited at geologically significant rates that are comparable to those for inorganic marine deposits.

In several instances marine biologists and geologists have measured the rate at which sediment is *re*-deposited by various marine organisms. Often these rates are of the same order of magnitude as primary rates of inorganic and organic deposition. For example, John Bardach, a fisheries biologist, found that many of the reef-dwelling fish of Bermuda (parrotfish, triggerfish, puffers, and the like) browse on the calcareous sand of the reef for its included food and for use as a milling agent in grinding the algae they have eaten. Bardach calculated that these fish annually redeposit between two and three metric tons of calcareous material per hectare (about one ton per acre) on a typical Bermuda reef. These data convert to a sedimentation rate of 10 to 15 centimeters per 1,000 years. Similar observations by Preston E. Cloud, Jr., who studied some Pacific reefs, yielded redeposition rates by fish of 20 to 30 centimeters per 1,000 years.

Burrowing marine invertebrates can also redistribute and redeposit materials after their initial deposition. Various studies—for instance, of worms, holothurians (sea cucumbers), and clams—indicate that these organisms are capable of reworking sediment just as fast as it is deposited by inorganic agents.

Table 2-2

Estimates of Marine Sedimentation Rates*

Carbonate sediments	Area	Geologic Age	Absolute Age (in Years)	Estimated Sedimentation Rates (cms/1.000 Years)
Shallow water	Bahamas	Late Cretaceous-Recent	100 million	4.5
" "	"	Last glaciation-Recent	11,000	38
Deep water	Caribbean	Pleistocene-Recent	400,000	2.5
Noncarbonate sediments				
Shallow water	Gulf Coast	Paleocene-Recent	70 million	20.0
Deep water	Various			0.05 to 0.5

From N. D. Newell, 1955, *Geol. Soc. America Spec. Paper 62*, p. 303; P. E. Cloud, Jr., 1956, *U.S.G.S. Prof. Paper 280-K*, p. 399; D. Ericson and G. Wollin, 1964, *The Deep and the Past*, Knopf, N.Y., p. 71; G. M. Kay and E. H. Colbert, 1965, *Stratigraphy and Life History*, Wiley, N.Y., p. 548; G. Arrhenius, 1963, *The Sea*, Interscience, N.Y., v. 3, p. 656.

* These rates are at best educated guesses and will vary from place to place. They are shown to give a general idea of relative sedimentation rates.

Sediments and environments

It seems clear, then, that organisms do make truly significant contributions to the sedimentary rock record through the formation of skeletal sedimentary grains as well as through the reworking and redeposition of sediment. In fact, when we once realize how effective organisms can be as rock-builders and sediment-modifiers, we may ask why *all* sedimentary rocks don't show these organic influences.

Some sedimentary rocks, such as evaporites, glacial deposits, and dune sandstones, will, of course, lack traces of any organic activity because they are formed in environments where organisms are virtually excluded. Other depositional environments may contain rock-building and sediment-modifying organisms, but the rate of organic activities may be far less than the rate of sediment influx and accumulation, so that the sediments are buried more rapidly than they can be effectively reworked by organisms. The production of calcareous sediments—either as skeletal sands, shell beds, or reefs—requires that only limited amounts of noncarbonate land sediments come into the sedimentary basin so that the organically produced sediments are not diluted. Hence, preservation of either inorganically produced primary structures (cross-stratification, ripple marks, graded bedding, and so on) or organically produced structures (burrows, pellets, shell beds) depends on the relative rates at which the inorganic and organic processes are working. For example, Keith Chave, a geologist who has studied various aspects of recent carbonate sedimentation, suggests from preliminary data that the production of organic calcareous sediments today is relatively constant (6 to 30 centimeters per 1,000 years) and that what really controls the ultimate formation of calcareous sedimentary deposits is the rate of their dilution by noncarbonate sediments. Similarly, John Imbrie and Hugh Buchanan, in their study of structures in Bahaman calcareous sediments, found that in turbulent shoal areas along the margins of the Great Bahama Bank, the sediments retained their primary current stratification, because of the constant reworking by tidal flow across the shoals and because of the relatively few bottom-dwelling burrowing organisms that could disturb the sediments. On the other hand, the calcareous sediments from the interior lagoon of the Great Bahama Bank lacked any primary stratification and, in the absence of any regular, strong bottom currents that could restratify the sediments, were thoroughly stirred up by the abundant burrowing fauna found there (Fig. 2–10).

Stratigraphic Relations of Sedimentary Rocks

At the moment of deposition, sediments have areal dimensions that are, of course, closely related to the nature of the depositing medium. Stream and beach deposits are linear, running parallel to the direction of stream flow or surf action. Deep-sea muds and lagoon sands will be blanket-like and usually wide-

FIGURE 2–10 *Box cores from the uppermost 30 cm of calcareous sands from the Great Bahama Bank. The core on the left shows excellent primary current stratification as developed on the foreshore of a beach. The core on the right shows no primary stratification and is heavily burrowed by marine animals living in these lagoon sediments. (Courtesy John Imbrie, S.E.P.M. Special Publication No. 12.)*

spread. Reefs and evaporate deposits are often quite discontinuous or patchy in their occurrence, and so their distribution is localized and irregular.

With time, however, this two-dimensional geometry assumes a third dimension, because as the sediments continue to accumulate they develop significant thickness. In order for sediments to build up to any great thickness, the sedimentary basin itself must subside to accommodate the ever-increasing quantity of sediments. The weight of the sediments alone is not sufficient to cause long-continued subsidence, because the density of the sediments (about 2.03) is less than that of the rocks that lie below the floor of the basin (about 2.65). Thus, for every 1,000 meters of sediment deposited, the sedimentary basin will subside only a little more than 750 meters. Subsidence due only to the weight of loading is therefore limited; long-continued sedimentation results in total filling of the basin and the area of deposition will become an area of nondeposition or erosion.

Certain places in the world do, indeed, have extremely thick sedimentary rock sections—several kilometers or more—and these occur where the earth's crust has shown considerable instability. For example, the linear mountain systems of the world mark those places where quantities of sediment were deposited in subsiding sedimentary troughs, or *geosynclines*, often accompanied by volcanic activity. Eventually, the subsidence ceases (reasons for the subsidence are still not altogether clear) and the mass of sediments is slowly compressed and deformed. Changes in pressure and temperature accompanying subsidence and later compression often *metamorphose* the sedimentary rocks into slates, schists, and gneisses. The total complex of sedimentary, metamorphic, and volcanic rocks that develops within a geosynclinal terrain thus forms the geologic backbone of various mountain systems of the world (Fig. 2–11).

During the history of a depositional area, whether a geosyncline or a lesser sedimentary basin, the particular site of sedimentation may shift with time.

Sediments and environments

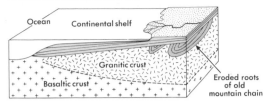

A. Sedimentation on continental border

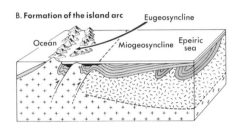

B. Formation of the island arc

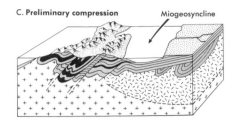

C. Preliminary compression

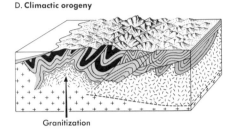

D. Climactic orogeny

FIGURE 2–11   *The history of a geosyncline. (A) Sediments accumulate along a continental margin—often an unstable area—as sediments are eroded from the older continental mass. (B) Instability of the area results in the formation of a volcanic island arc with adjacent sedimentary basins ("eugeosyncline") and further subsidence of the continental margin ("miogeosyncline") results in continued sedimentation and even flooding of the old continent by shallow seas ("epeiric"). (C) Compression accompanies continued arc formation with some metamorphism of the sediments and volcanics. (D) Subsidence stops and the whole geosynclinal complex is thrust toward the continental mass; rocks are metamorphosed and deformed ("orogenic climax"). Rocks at the bases of the geosyncline are extremely metamorphosed to the extent that they become viscous liquids and intrude surrounding rocks ("granitization"). The geosynclinal mass forms a thick prism of rocks which have considerable topographic relief ("mountain system") and which will themselves be gradually eroded away. (After Clark and Stearn, 1960.)*

Consequently, the position of a given sedimentary rock body may not be constant within a stratigraphic sequence (that is, the local sedimentary record) but instead will vary laterally and vertically, being the locus of successive depositional sites through time. For example, a river will often meander back and forth across its flood plain. The sedimentary record of this migration is recorded as lense-shaped bodies of channel sands and oxbow lake muds incorporated within interchannel silts (Fig. 2–12).

So far we have been discussing the sediments within a sedimentary basin as if they were all alike throughout the basin. Obviously, such is not the case. Within a basin of sedimentation there are, usually, a number of different, local depositional environments. These local environments reflect variations in physical, chemical, and biological conditions within the basin as well as their distance and direction from any depositional agent that may be entering the basin—such as a river with its associated delta. Hence, at any one time the sediments being deposited throughout the basin will have different characteristics and over-all aspects that are correlated to the local depositional environments. Such lateral

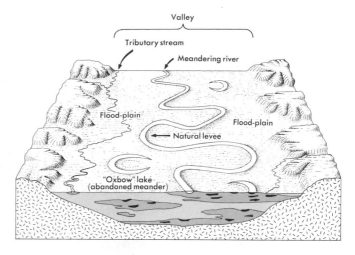

FIGURE 2–12 *Block diagram of the valley of a meandering river, showing a diagrammatic cross-section. Deposits (dark gray) occur as irregular lenses of sandstone within interchannel sediments (light gray), usually shales and siltstones. Cut-off meanders may form "oxbow" lakes where fine-grained muds accumulate (black). (After C. O. Dunbar and J. Rodgers, 1957.)*

variations within a sedimentary basin are termed *sedimentary facies* (Fig. 2–13). The depositional sites of these individual facies may also shift their position through time so that each facies will have its own three-dimensional configuration in the total stratigraphic sequence deposited within the basin as a whole (Fig. 2–13).

Earlier in this chapter we indicated how sediments will have inorganic and

FIGURE 2–13 *Cross section of the Mancos Shale and Mesaverde Group of eastern Utah and western Colorado. These Upper Cretaceous deposits record nonmarine and nearshore marine sediments (Mesaverde Group) interfingering eastward with offshore marine deposits (Mancos Shale). The stratigraphic section as a whole indicates gradual withdrawal of the sea to the east as erosional debris enters the area from the west during the uplift of the ancestral Rocky Mountains. Note the complex intertonguing of different sedimentary rock types, which records constant shifting of local depositional environments with time. (After C. O. Dunbar and J. Rodgers, 1957.)*

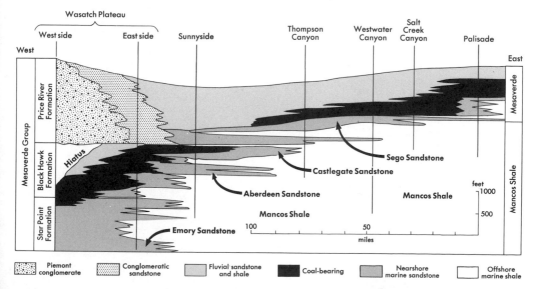

organic attributes of texture, composition, internal structures, fossils, and so on. Correspondingly, sedimentary facies may be characterized from their inorganic, lithologic qualities—"*lithofacies*"—as well as from their biologic qualities—"*biofacies*." In either case, lithofacies and biofacies are a direct manifestation of the local depositional environment within a sedimentary basin. In order to define and interpret the origin and history of a given stratigraphic sequence, it is useful to identify first the various sedimentary facies contained therein. Instead of speaking of "20 meters of upper Cretaceous sandstone in this county and 50 meters of shale in the next" we may say "nearshore quartz sands here, passing into offshore shallow marine muds over there." The recognition and interpretation of facies thus permit us to move from a purely descriptive narrative of what is seen in scattered rock outcrops (certainly a necessary first step) to a discussion of how and why certain kinds of rocks accumulated where they

FIGURE 2–14   *Detailed analysis of the Pocono Formation, Mississippian, of the central Appalachians. This study by B. Pelletier included many aspects of this sedimentary rock, including grain size and composition, orientation of cross-stratification and plant remains, maximum size of quartz pebbles, sand/shale ratios, and fossil content. Pelletier was able to demonstrate that the Pocono Formation was a nonmarine, coastal-plain sediment derived from sedimentary rocks and low-grade metamorphic rocks in a source area located near Atlantic City, New Jersey. Sediment transport was to the west and northwest; the ancient shoreline trended northeast across Pennsylvania and was located some 25 miles east of Pittsburgh. The offshore facies of the Pocono is a marine shale and sandstone that contains abundant burrows ("Arthrophycus"), occasional brachiopods, and a few clams and snails. Sand/shale ratio greater than two is shaded; maximum pebble diameters in millimeters shown by contours; current directions by small arrows; note relation of oil pools (black) to facies. (After B. Pelletier, 1958.)*

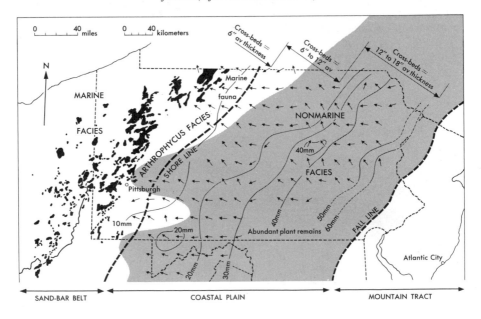

did, what their relationships are to each other, and what sorts of life habitats might have existed for any associated organisms.

Summary

Throughout this chapter the importance of the local environment in controlling the genesis of a sedimentary rock has been stressed. Thus, the composition of the source rock and the rate at which it is weathered influences the composition of the sedimentary grains brought to the basin of deposition. The nature of the transporting medium and the conditions existing during and after sediment deposition control the texture and primary structures of the sediments. The postdepositional environment determines the degree and kind of consolidation and lithification of the sediments into rocks. Organisms, too, contribute grains in the form of skeletal debris to the sediments and the presence and distribution of organisms, in turn, is determined by the local environment. Organisms, moreover, can rework sediments to such an extent that their original textures and structures are significantly altered. Finally, the over-all structural and sedimentary framework limits the lateral extent and vertical thickness of a sedimentary unit as well as any facies development within. In short, sediments have complex and changing geologic environments that leave their superposed traces on the resultant sedimentary rock. Careful analysis, however, of the stratigraphic relations, facies, composition, textures, structures, and fossils may reveal the nature and sequence of these individual environments (Fig. 2–14).

3

Organisms and environments

Adaptive Responses of Organisms

Ordinarily we do not think about the environmental adjustments that animals and plants have made during their evolutionary history. The reason is, of course, obvious. Since human beings are constantly manipulating their environment so that we are able to penetrate and colonize the most hostile surroundings (for example, Antarctica, deserts, and, presumably, outer space), we have lost an awareness of the strong control that environment usually exerts in limiting the distribution and abundance of animals and plants.

It is a fact, certainly, that organisms are heavily dependent on the surrounding external world for satisfying their basic needs for living space, food, and mates. And because organisms have varied in their success in adjusting genetically to the local environment, they have had varied success in leaving offspring. Such differential reproductive success results in a gradual change or shift in the genetic background of a species. This change is usually in the direction of increased adaptation to the environment so that with the passage of generations, the species eventually optimizes its total genetic complex with respect to the local environment. As long as the environment remains the same, further changes within the species are selected against and the species remains at its adaptive position; significant changes in environment will require new adaptations.

There is another, related way in which organisms interact with their environment. The genetic background, or *genotype*, of an individual organism establishes a *range of reaction* of the individual to its environment, and its *particular* expression is influenced by the specific environment in which it finds itself. The resulting, visible product of this interaction between the genotype and the environment is what geneticists term the *phenotype*. Thus, for example, the *potential* size that a given human being may attain is controlled by his genotype. The *actual* body size that each of us attains is the result of the interaction of our individual genotypes with our individual environments. Differences in the body size of human beings are, therefore, phenotypic expressions of different human genotypes interacting with different environments.

The great diversity that we see in nature is the evolutionary result of many different genetic types invading the wide variety of environments found on our planet. These environments, or *habitats*, are extremely varied and include the prevailing physical and chemical factors, such as temperature, humidity, salinity, content of oxygen and carbon dioxide, and light. Equally important these habitats also include biological factors, such as potential competitors for food and living space, predators and prey, parasites and disease-causing micro-organisms, and density of numbers of members of the same species. These habitats may vary from lush, tropical, rain-forests to hot, arid, treeless plateaus; from cold, dark, organic-rich muds of the deep sea to the warm, well-lighted, agitated waters of a coral reef; from grassy, wide-ranging plains to the dark, moist, and convoluted lining of the mammalian intestine. Within each of these diverse habitats there are a number of *niches* that many kinds of organisms may fill.

Charles Elton, a long-time student of environments and their occupants, has described the habitat as the species' "address," and the niche as the way the species makes its "living" at that address. Continuing the analogy, Elton refers to the typical, rural English village (habitat or address) with its various niches filled by the village vicar, chemist, solicitor, and doctor. Similarly, in the coral-reef habitat there is a multitude of niches ranging from that of the coral polyps sitting in their calcareous colonies and feeding on minute animals suspended in the surrounding sea water to that of burrowing sea cucumbers that eat the bottom sediment for its included organic matter to the predatory barracuda that cruises continuously in search of an unwary victim.

The success of an organism in its niche within a particular habitat is measured by its over-all adaptation to that niche and the relative number of offspring it leaves as a consequence of that adaptation. The adaptation of an organism expresses itself in myriad ways, including a variety of morphologic, physiologic, behavioral, reproductive, and developmental characteristics and mechanisms necessary to cope with the environment. The pigments in plants, for example, are so colored as to receive particular portions of the available spectrum of sunlight for maximum photosynthetic efficiency. Some flowering plants have evolved a flower structure that, by attracting insects with its nectar, ensures the dusting of an insect with pollen (the male germ cells). When the insect visits another plant of the same species, this pollen is transmitted, thereby fertilizing the second plant. Insect-flower relationships have great reproductive benefits for plants, which are immobile and so cannot seek out a mate. The insects also benefit, of course, by exploiting the food provided by the nectar.

Other types of adaptations are related to the skeletal parts of an organism—wings for flying, fins for swimming, teeth for cutting and chewing food, shell ribbing for burrowing. Consequently, the structure of such parts contains considerable information about the owner's way of life.

Because such skeletal parts are often composed of secreted crystalline materials—bone, shell or teeth—which are relatively resistant to mechanical disintegration and chemical and bacterial decomposition, they are commonly fossilized. For this reason, then, students of ancient organisms and environments pay particular attention to the structural, or morphological, relations of fossilized skeletal parts in order to make inferences about the habitat and habits of the organism. This emphasis by paleontologists on *adaptive morphology* in fossils is necessary because other adaptive characteristics—whether physiological, behavioral, reproductive, or developmental—are almost always "soft-part" features and hence are rarely preserved in the fossil state. Since protoplasm, tissues, and various organs are completely without preservable hard, crystalline material, they are usually soon disintegrated, decomposed, or eaten after the death of an organism so that burial within sediments and subsequent fossilization is very remote.

It is unfortunately true, therefore, that certain adaptive aspects of ancient organisms will be forever unknown and unknowable. Yet the fossil record is sufficiently adequate to arrive at least at first conclusions regarding the nature of many ancient environments and the roles that various fossil forms played in those environments.

Functional Morphology

For an organism to survive and reproduce, it must perform a number of specific functions repeatedly and effectively throughout its life. These functions include, among other things, the searching out of food, shelter, and mates. Food must be gathered and incorporated within the organism where it can be digested and assimilated. This is as true for the elm tree that derives its water and nutrients from the soil through its many thousands of root tips as it is for the sponge that circulates water through its elaborate body-canal system, where millions of small cells filter out the suspended food particles.

Organisms must also establish themselves in an environment where they can survive in the local physical, chemical, and biological conditions. Therefore, there must be some mechanism for moving about in the general area and for "settling in" once a favorable site is located. For example, many marine invertebrates, although as adults they may be sedentary, bottom-dwelling forms, have a free-floating period in their early life, or *larval stage,* during which the larvae are dispersed far and wide by the ocean currents. After an interval of time ranging from hours to weeks, the larvae settle to the sea floor and undergo a metamorphosis to a miniature adult. During their free-floating stage, the larvae are usually equipped with some specific morphologic character, such as tufts and bands of cilia, that aid in keeping them suspended in the water and in generating small currents around the mouth for filter-feeding. After metamorphosis, other morphologic structures develop that enable the organism to burrow into the sediment (the foot of some clams), to crawl along the sediment (the various appendages of a lobster), or to secrete a stony, external skeleton that is fixed directly to the sea-floor (the calcium-carbonate-secreting tissue of certain corals).

Survival of the individual, important as that is, does not necessarily ensure the life of the species. Therefore, organisms must reproduce themselves if the race is to continue. Here, too, one can discover a great variety of morphologic features—as well as physiologic and behavioral mechanisms—that are adapted for the production and fertilization of germ cells, or *gametes,* and, in some instances, for care of the young. The songs, plumage, and nesting-habits of birds are all part of an elaborate reproductive complex to attract mates, effect fertilization, and care for the eggs and hatched young.

So we see, then, that organisms have a variety of functions to perform including feeding, locomotion, and reproduction if they and their species are to survive. Now, many of these functions are closely correlated to specific, hard skeletal parts of the organism. If such hard parts are found preserved in the fossil state, then it should be possible to infer the function that such a part played during the life of its owner. This, in turn, should tell us something of the environment where that organism lived.

For example, the skeletal appendages of various vertebrates are specifically adapted—through a long period of natural selection and evolution, of course—for locomotion in water, on land, or in the air (Fig. 3–1). Fish have short and broad appendages with a relatively large surface area for paddling through the dense medium of water. Land-dwelling vertebrates have longer and narrower limbs for supporting the body in walking and running. In the case of those groups that have returned to the sea (ichthyosauran reptiles and cetacean mammals), the limbs have been modified once more and are stubbier and more spatulate than before. Flying vertebrates, the birds and bats, have a limb that, when covered with feathers or a skin membrane, forms an aerodynamically stable structure.

Another example, again from the vertebrates, of the close correlation be-

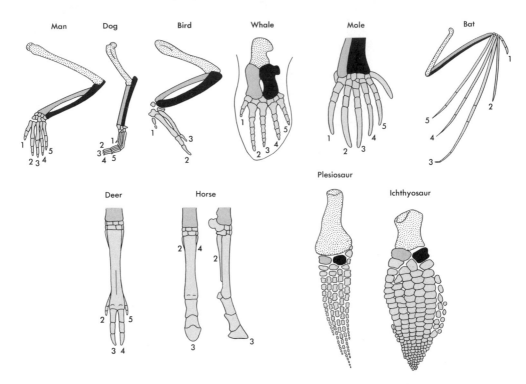

FIGURE 3–1   *Structure of various vertebrate forelimbs showing the arrangement of homologous bones. Each limb is specifically adapted for a particular way of life, whether burrowing (mole), flying (bat, bird), swimming (whale, ichthyosaur, plesiosaur), running (dog, deer, horse), or manipulating objects and tools (man). The various forelimb bones include: humerus, stippled; radius, dark gray; ulna, black; metacarpals (wrist bones) and phalanges (finger bones), light gray; numbers refer to homologous digits (fingers).*

tween the form and function of a morphologic character is provided by mammalian teeth. Every species of mammal, living or extinct, has a unique set of dental characters that are directly related to its diet.

The basic dental formula in mammals consists of 44 teeth that can be subdivided into four sets of 11 homologous teeth: left upper, left lower, right upper, and right lower. In the less-specialized mammals these 11 teeth are, progressing from front to back, three incisors for nipping, one canine for tearing or stabbing, four premolars for cutting, and three molars for chewing or grinding. The differences in the shape and size of these teeth reflect a division of masticatory labor among the teeth in biting, cutting, and chewing the food (for three widely varying examples, see Fig. 3–2).

Many mammals deviate from this general plan because of specialized diets. The carnivores have particularly well-developed canines for stabbing as well as modified first lower molars and last upper premolars, which have become sharp, crested shearing blades for cutting—two adaptations well-suited for the eating of freshly killed meat. In contrast, the vegetarian herbivores such as cattle have lost their canines while their premolars have been modified into molar-like teeth suitable for grinding and chewing tough vegetable matter.

Because mammal teeth are composed largely of crystalline calcium phosphate, they are mechanically strong and chemically stable and so are commonly found as fossils in sedimentary rocks. Furthermore, there is a close interdependence between dentition, diet, and way of life. For these two reasons, then, students of mammalian evolution have been able to document and unravel the complex evolutionary history of mammals for the last 100 million years and more, basing much of their interpretation on fossilized teeth.

It is not, however, always easy to demonstrate the functional significance of a particular hard part, principally because many fossil organisms are extinct and lack living, flesh-and-blood, representatives that can be observed in the field and dissected in the laboratory. It is often impossible to establish by analogy what function a given hard part served during the life of its owner. Such an example of well-known but poorly understood hard-part morphology is provided by an extinct group of shell-bearing protozoans called *fusulinids*, which lived some 225 to 325 million years ago during the late Paleozoic. During the 100 million years of their existence, the fusulinids underwent considerable evolution, much of which is recorded in various morphologic features of the small, calcareous shell (or test) that they secreted. Some of these changes included overall increase in size, variation in test shape, rotation of the axis of coiling from the shortest diameter to the longest diameter, amount and position of secondary calcium carbonate deposits within the axial portion of the test, and an increase in the fluting of the partitions that internally subdivided the shell. The evolution

FIGURE 3–2   (A) Dentition of a relatively unspecialized mammal showing three incisors (nipping and biting), one canine (stabbing and tearing), four premolars (cutting), and three molars (grinding and chewing) on each half of the upper and lower jaw. (B) Two highly specialized mammals with dentition which is considerably different from that illustrated in (A): the extinct Pleistocene saber-tooth cat on the left and the modern cow on the right. The saber-tooth cat was a predaceous carnivore equipped with large, stabbing canines and modified upper premolars and lower molars ("carnassial teeth") for cutting up its freshly caught meal. The herbivorous cow has lost its upper incisors and canines; it crops its food of grasses and small plants with the projecting lower incisors that press against a tough, horny pad on the upper jaw. Since a cow's premolars are modified into molar-like teeth, the animal is thus provided with a battery of grinding teeth for thoroughly grinding its food. (From Romer, 1962.)

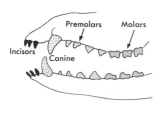

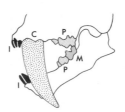

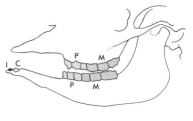

A

B

of these various characters resulted in the rise within this single family of protozoans of some 40 to 50 genera with 1,000 species. And yet, despite the precision
with which these evolutionary trends in different morphologic characters can
be documented, the functional and adaptive significance of these characters is
not at all clear.

In short, therefore, various morphologic aspects of fossil organisms often
reveal the habitat and habits of the organism itself as well as contribute information about the ancient environment where that organism lived. But the
code relating form to function has not been solved for many, perhaps even
the majority, of fossil organisms, particularly those that are extinct and are
without a closely related living descendant.

Some Important Environmental Factors: Physical and Chemical

Because of the various specific adaptations organisms have made to particular habitats during their evolutionary history, they are usually limited to
those habitats where the local environmental conditions are favorable. These
conditions include a complex array of physical, chemical, and biological phenomena that are not only important by themselves but also in their mutual
interaction in limiting the distribution and abundance of a specific group of
animals or plants. We will review some of the principal ecologic factors that
exert an influence on various kinds of organisms. This review, besides being
somewhat general and necessarily incomplete, will treat these individual
ecologic factors as if they were acting independently; but remember that these
factors—as well as others omitted from our discussion—are usually interacting
with one another, and that the over-all impact of the local environment is
greater than the sum of its individual aspects.

Temperature

One of the important factors of any environment is the *temperature* of the
surrounding medium, whether water, air or soil. The reason is that the
temperature of an organism's surroundings strongly influences its internal temperature in most instances—the most obvious exceptions being birds and mammals, which maintain relatively precise internal temperatures irrespective of
the external temperature (within limits, of course).

As the internal temperature of an organism varies—and it varies more or
less as the external temperature varies—reaction rates of many physiological
processes will also vary because these reactions are mainly chemical in nature
and are, therefore, speeded up or slowed down with an increase or decrease in
the local external temperature. As a result, rates of metabolism, development,

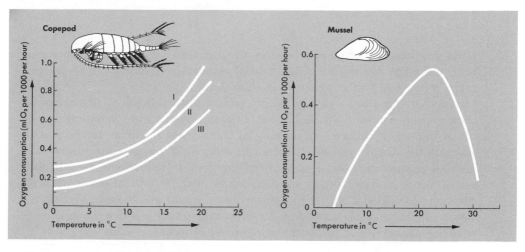

FIGURE 3–3 *Graph on left shows rapidly increasing respiration rates in male (I), female (II), and juvenile (III) copepods,* Calanus finmarchius, *as temperature of the sea water increases. Graph on right shows similar respiratory increase in the common mussel,* Mytilus edulis. *Notice, however, that above 22°C, respiration rates decline sharply owing to the general debilitating effect of higher temperatures on these organisms. (From H. B. Moore, 1958.)*

growth, and reproduction will ordinarily fluctuate with temperature changes. This change of reaction rates with temperature change is summarized by *Van't Hoff's rule,* which states that for every 10°C of temperature increase, organic reaction rates increase by a factor of 2 (probably more accurately by a factor of 1 to 6). Although this rule cannot be universally applied because of many exceptions, it does suggest the order of magnitude of temperature control on many organisms (Figs. 3–3 and 3–4).

Besides these direct influences on physiological reaction rates, temperature variations have other, less direct, effects that may be of equal significance. For example, the solubility of various solids and gases that are critical in supporting life, such as nitrates, phosphates, oxygen, and carbon dioxide, vary with tem-

FIGURE 3–4 *Increasing growth rate (during one month) with temperature increase in the oyster,* Crassostrea virginica. *At 15°C there is an optimum temperature for growth; above this value, growth rates lessen. (After H. B. Moore, 1958.)*

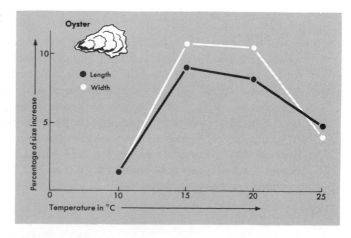

perature, the solids being more soluble, the gases less soluble, with an increase in temperature. The availability of these substances to organisms will, therefore, often be related to the surrounding temperature.

Another important way in which temperature influences organisms is through convection of air and water. Periodic temperature variations bring about density differences within air and water masses. Falling temperatures increase the density causing the mass to sink, while rising temperatures will cause it to rise. Such convective movements tend to redistribute and mix the air or water and so increase the circulation of other critical environmental elements contained within the mass, such as nutrients and oxygen. This circulation can, for example, ameliorate an environment that was deteriorating during the stagnant period before the convective circulation.

Oxygen and Carbon Dioxide

Two atmospheric gases, carbon dioxide and oxygen, are especially important to animals and plants. Oxygen is, of course, critical for all organisms because of its role in cellular respiration. Its importance derives from the fact that when various organic molecules such as amino acids, proteins, carbohydrates, and starches, which are either produced by the organism or else obtained from its food, are oxidized, they thereby release their bound-up chemical energy. This energy, which is part of the solar radiant energy originally stored in the organic molecules that are manufactured by plants during photosynthesis, is released in turn in various forms, including those of thermal energy, kinetic energy, and electrical energy, all of which the organism needs to sustain life.

This storage of chemical energy in organic molecules by plants during photosynthesis, and its release by plants and animals during cellular respiration, is summarized by the general equation:

$$\underset{\substack{\text{Solar} \\ \text{radiation}}}{\text{Energy}} + \underset{\substack{\text{Carbon} \\ \text{dioxide}}}{CO_2} + \underset{\text{Water}}{H_2O} \underset{\text{respiration}}{\overset{\text{photosynthesis}}{\rightleftharpoons}} \underset{\text{Sugar}}{C_6H_{12}O_6} + \underset{\text{Oxygen}}{O_2}$$

As this equation states, carbon dioxide is as important to biological processes as oxygen because it is a basic building block for organic matter—either for food or for the formation of new protoplasm.

Although there are some organisms—anaerobic bacteria, in particular—that can liberate energy from organic molecules without oxygen, most organisms need oxygen for cellular respiration. (Some of these, too, can liberate oxygen from organic molecules without oxygen, but only for limited periods of time because they accumulate an oxygen debt which, like all debts, eventually becomes due.) Oxygen is obtained either from the atmosphere or from solution in water. Oxygen constitutes about 21 per cent of the volume of the atmosphere and is normally sufficiently abundant to meet the requirements of all air-breathing organisms. Aquatic organisms, on the other hand, have a smaller and more variable oxygen supply, since the amount of oxygen dissolved in water varies directly with temperature and inversely with salinity of the water. The oxygen content also fluctuates because it is added by plant photosynthesis and subtracted by organic respiration and decomposition (oxidation) of organic matter. As a result, fresh water—lakes, ponds, and streams—and sea water usually contain less than 1 per cent oxygen by volume. In both cases, oxygen may be totally depleted in various stagnant, aquatic environments (for instance, in depths greater than 200 meters in the Black Sea, or at the bottom of deep, narrow fjords, certain swamps, and deep lakes) which, except for anaerobic bacteria, will be completely devoid of all life.

Carbon dioxide, besides being necessary for photosynthetic organisms, is also ecologically important because it influences the pH, or hydrogen-ion concentration, of natural waters. When atmospheric or oxidative carbon dioxide dissolves in water, it also forms a weak acid, H_2CO_3, which dissociates to form hydrogen ions and bicarbonate ions. The hydrogen-ion concentration, or acidity, of the water, therefore, increases with increasing carbon dioxide in solution. The production of the bicarbonate ion may become important in buffering water against sudden changes in acidity or alkalinity.

In the sea, especially, the carbon dioxide content may play a significant role in the solubility of calcium carbonate and, in turn, influence its availability for calcium-carbonate-secreting organisms. In the deeper parts of the oceans the waters are relatively cold and under great pressure; they can, therefore, hold more carbon dioxide in solution than the warmer, surface waters. At these depths there is solution of the many millions of microscopic, calcareous skeletons secreted by planktonic protozoans in the upper, surface waters. Thus, it happens that in oceanic depths greater than 5,000 meters these skeletons have all virtually dissolved away while falling downward; at shallower depths, however, they become increasingly abundant in the bottom sediments and form, in many places, a fine-grained calcareous sediment.

Sunlight

Solar radiation, or light, is another important ecologic factor because it is the basic energy source for *all* organic activity. As indicated earlier, plants convert carbon dioxide and water during photosynthesis into various organic molecules—particularly carbohydrates, as well as fatty acids and proteins—which, in turn, provide a food source not only for the plants themselves but also for all animals, which, unlike plants, cannot carry out this conversion. Therefore, it is through the intermediary process of plant photosynthesis that all life ultimately depends on solar radiant energy.

The amount and duration of solar radiation vary greatly throughout the many geographic areas and environments found on the Earth. Because the Earth's axis of daily rotation is inclined to its axis of yearly revolution about the sun, the solar energy received by various parts of the planet changes, thereby creating day-night as well as seasonal cycles. These seasonal variations in solar radiation are greater the farther one goes from the equator. Whereas plants in the tropics receive strong light for about 12 hours of every day of the year, plants in the arctic regions receive radiation that is much less intense and very unevenly distributed. In fact, in very high latitudes there is continuous daylight in the summer and continuous darkness during the winter.

There is also conspicuous variation in the amount of solar energy received by the oceans and land-locked bodies of water. Besides the latitudinal variations, the depth of penetration of light down into the water is limited and rarely goes beyond about 200 meters below the water's surface. This zone of maximum light penetration is referred to as the *photic zone*. Not only does the amount of radiation decrease with water depth, but the wave length, or color, of the light also changes. The less intense, longer-wave-length red end of the visible spectrum—as well as the infra-red, or heat, portion—is more readily absorbed by the water than is the more intense, shorter-wave-length blue end. It is partly for this reason that there is a general change in the pigmentation of marine algae with water depth. Green algae commonly occur in shallow waters, brown algae in deeper waters, and red algae at the greatest depth. The characteristic depths of these various marine algae reflect adaptations in the development of different colored pigments that are *complementary* to the wave length (or color) of the light available at these different water depths. There may be other adaptations as well, such as increasing the concentration of photosynthetic pigments to compensate for the decrease of radiation of a particular wave length.

In the *aphotic* regions of the oceans or deep lakes where no light is available, animals must depend for their food supply on the rain of organic matter that comes down from the photic zone. Some animals, of course, prey upon other animals that feed on this drifting organic detritus.

Salinity

In aquatic environments the total amount of dissolved solids, or salinity, is an important environmental factor. Terrestrial waters—streams and rivers, ponds and lakes—differ from oceanic waters in that they are less saline, their salinity is more variable, and they generally contain other proportions of salts. Thus, although the oceans usually contain 3 to 4 per cent of salts in solution, river waters vary from less than .001 to almost 1 per cent (varying, therefore, by a factor of 1,000) and average about .01 salinity. The bulk (about 99 per cent) of the substances in solution in sea water are, in decreasing abundance, chloride, sodium, sulfate, magnesium, and potassium ions. In river water, on the other hand, the same percentage of dissolved materials is composed of carbonate, calcium, sulfate, silica, sodium, chloride, magnesium, iron and aluminum oxides, and potassium. In sea water the major dissolved constituent is sodium chloride—

hence, the "salty" taste of the oceans—and in river water it is calcium carbonate—which forms the scale in tea kettles.

Note the apparent discrepancy here. If the rivers are carrying mostly calcium carbonate to the oceans, why is sodium chloride the major oceanic salt? Chiefly because many marine organisms whose skeletons are made of calcium carbonate are actively extracting large quantities from sea water. Much of this calcium carbonate is buried in marine sediments as fossil debris. If and when these marine sediments are uplifted and eroded, this bound-up calcium carbonate will be carried back to the sea. Sodium chloride, by contrast, is retained in solution, and although it is added to the oceans at a slower rate than calcium carbonate, it has had a greater net accumulation during geologic time.

The single greatest effect of salinity on aquatic organisms is *osmotic pressure*. The cells of organisms are essentially viscous chemical solutions held together by a membrane which, because its function is to keep the cell contents intact, is only partly permeable, allowing just water and certain ions of small diameter to pass back and forth. This movement of water between cells and the exterior is called *osmosis*. When a cell is in a watery medium with a salinity different from that of the cell fluids, the water pressure is directly proportional to the difference in salinity. If the cell's salinity is less than that of the external medium, then water tends to move out from the cell; if the cell salinity is the greater, then water tends to move into the cell. Unless the osmotic pressure and water flow are regulated by the organism, this process of osmosis will result in either the desiccation of the cell through excessive water loss or the flooding of the cell until it eventually bursts.

Many organisms have evolved osmoregulatory mechanisms either for conserving water or for removing excess fluids; such mechanisms allow the organism to tolerate salinity fluctuations in the environment. However, because of the great differences in quantity and composition of the substances dissolved in terrestrial and marine waters, many organisms that originated in the sea have never been able to succeed in colonizing terrestrial waters (for example, echinoderms and cephalopods). Other groups, such as fish, that are able to inhabit both fresh and sea waters, have radically different osmoregulation (Fig. 3–5).

Water Turbulence

Water energy, or turbulence, is another important ecologic factor in aquatic environments, especially the shallower parts of the oceans and lakes. Water turbulence helps to ensure an even distribution of food, nutrient elements, oxygen, and carbon dioxide throughout the environment as well as to remove toxic waste products. Consequently, this mixing of the water may prevent a local deterioration of the environment. Water turbulence also disperses the larvae of sedentary invertebrates, thereby helping maintain the areal distribution of these species.

Too turbulent an environment, however, is unfavorable if sedimentary

Organisms and environments

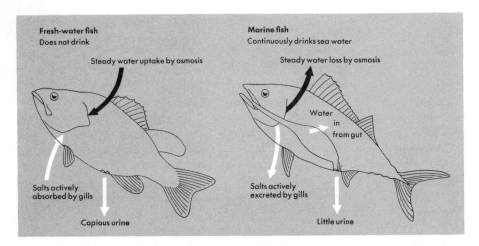

Figure 3–5 *Different mechanisms to maintain proper water balance of the body fluids in fresh-water fishes and marine fishes. In fresh-water fishes the salt concentration of the cells and tissues is greater than that of the surrounding water. There is a constant absorption therefore, of water by these fish; most of this excess water is removed by a heavy urine flow. To counteract the loss of certain needed salts dissolved in the urine, the kidneys of fresh-water fish reabsorb a large fraction of these salts from the urine before it is finally excreted. They can also absorb these necessary salts from the water with their gills. Marine fish, on the other hand, have a lower salt concentration in their body fluids relative to the surrounding sea water. To prevent water loss through osmosis they excrete very little urine and drink large quantities of sea water. However, to avoid further upset of their water balance, they secrete through their gills the salts which were dissolved in the sea water that they have drunk. (After Simpson and Beck, 1965.)*

grains are transported along the bottom in such a way that burrowing organisms are uncovered or organisms sitting on, or attached to, the substrate are buried. If the water energy is too vigorous, bottom dwellers may also be uprooted, dislodged, or overturned.

Water turbulence determines the size of grains of sediment that accumulate in a given area. The smaller grains settle out in quieter water, whereas the larger grains are carried in suspension or moved along the bottom by traction in more turbulent water. Thus, water energy indirectly influences other characteristics of a sediment related to grain size, such as content of organic matter (an important food source for many organisms), porosity, permeability, and sorting. These characteristics, in turn, control the ease of burrowing into the sediment, water content, and interstitial space for minute sediment-dwellers.

Marine biologists have suggested from time to time that the variations in the shape of certain shallow-water species of snails, sea urchins, and corals are related to differing degrees of wave exposure. For example, the stinging hydrozoan coral, *Millepora*, grows as an encrusting, hummocky colony in very shallow tropical waters. In increasing water depths, where turbulence is reduced, the colonies assume a vertically bladed, labyrinthine shape. Such changes in the gross form of an organism might be useful in interpreting the amount of wave energy or exposure in a fossil environment. Unfortunately, there is not as yet a sufficient body of reliable data relating the gross morphology of an individual organism or colony of organisms to their habitat and, therefore, any such interpretations from the fossil record must be regarded cautiously.

Substrate

Most organisms, even aerial forms, are not suspended in "thin air" but must have a surface, or substrate, upon which they pass their lives feeding, seeking mates, avoiding predators, and resting. In fact, many organisms are substrate-specific, that is, they will search out, and only survive upon, a surface having certain characteristics of background color, food content, firmness, composition, and so on. We know, for example, that many plants will thrive only in soils which have just the right nitrogen content, moisture, and granular texture. Some marine larvae can delay metamorphosis after once settling on the sea floor if the substrate is not suitable; this delay may increase their chances of being dispersed to a more favorable sediment type elsewhere. Certain aquatic species, such as clams and worms, burrow into the substrate for shelter and food.

There are some organisms that prefer a substrate of particular color because they may be virtually invisible there. Some animals, such as the flounder and chameleon, can modify their skin color so that they blend in more harmoniously with their background. The prime advantage of blending in with the substrate is, of course, avoiding or hiding from predators.

An interesting example of the important relation between organism and substrate is provided by a species of moth found in Great Britain. In rural areas today, and throughout the British Isles before the Industrial Revolution, this moth had a "salt and pepper" coloration, that is, dark speckles on a lighter background. Such coloration was obviously adaptive because this particular species of moth spends the daylight hours sitting on light-colored tree trunks and lichen-covered rocks. Against such a background these moths are practically invisible and therefore overlooked by preying birds. With the onset of the Industrial Revolution, however, the trees and rocks in urban areas became very dark, almost black, and many of the light-colored lichens were killed by the fallout of soot, ashes, and other pollutants from factories and mills. Today, except in rural areas where the air is uncontaminated by this industrial fallout, populations of these moths are dominated by a dark variety which is hardly discernible against the soot-covered rocks and trees in areas near the great industrial cities. This dark variety of moth is a mutant strain within the species that has become selectively favored in urban areas of Great Britain since the Industrial Revolution. It is clear, then, that the changing proportions of light and dark varieties of this moth correlate to the color of the background on which they rest during the day (Fig. 3–6).

Organisms and environments

FIGURE 3–6   *Light and dark varieties of a British moth showing relative inconspicuousness of each on a sooty tree trunk and on a lichen-covered tree trunk. (Courtesy H. B. D. Kettlewell, Oxford University.)*

In the marine realm, the substrate also exerts an influence on organisms. Muddy sediments are usually colonized by burrowing invertebrates that can easily penetrate these soft and watery substrates for shelter and food—either fine-grained organic detritus or bacteria and other micro-organisms. Coarser-grained sediments are more difficult to burrow into because the grains are usually well sorted, permitting close packing, and the water content is less. Consequently such sediments usually have fewer burrowing organisms than do muddier, finer-grained sediments.

Of all the physical-chemical factors we have discussed, the substrate is especially valuable in understanding *ancient* environments. The reasons for this are as follows.

1. Normally, the substrate of the environment becomes the sedimentary rock matrix with little postdepositional alteration beyond compaction and cementation.

2. Other physical-chemical factors, such as temperature, salinity, and oxygen and nutrient content, are usually *not directly recorded* within the sediments.

Organisms and environments

3. The substrate is a result of the complex interaction of the topography, hydrography, and current regime of the local sedimentary basin in which the sediments are accumulating and the organisms are living.

4. The organisms, particularly the bottom-dwelling forms, are influenced both by the substrate itself and by those other environmental factors associated with a given substrate.

Figure 3–7 illustrates these relationships of the substrate with other ecologic variables as well as the over-all control of the sedimentary environment by the topography, hydrography, and water energy of the local basin of sediment accumulation. This diagram is the interpretive result of studies, made by a group of paleontologists from Columbia University, of the dominant ecological

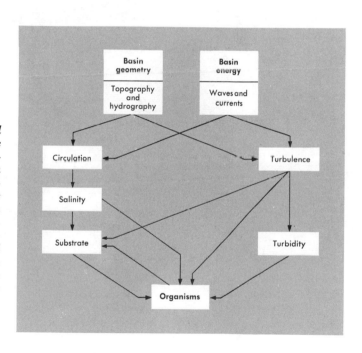

FIGURE 3–7 *The inferred relationships among the major environmental variables of the Great Bahama Bank in determining the nature of the substrate and the abundance and composition of the bottom-dwelling invertebrates. Note the fundamental dependence of all these ecologic factors upon the geometry and water energy of this shallow, subtropical marine environment. (From N. D. Newell and others, 1959.)*

factors responsible for the nature and distribution of sediment types and bottom communities on the Great Bahama Bank. Figure 3–8 shows the close correlation between sediments and bottom communities in this shallow, subtropical, marine environment. The reason for this close correlation is that the composition of the bottom-dwelling invertebrate communities is strongly influenced not only by the substrate itself, but also by the ecologic factors responsible for a particular substrate type. These associated factors are chiefly water turbulence, salinity, and circulation.

Organisms and environments

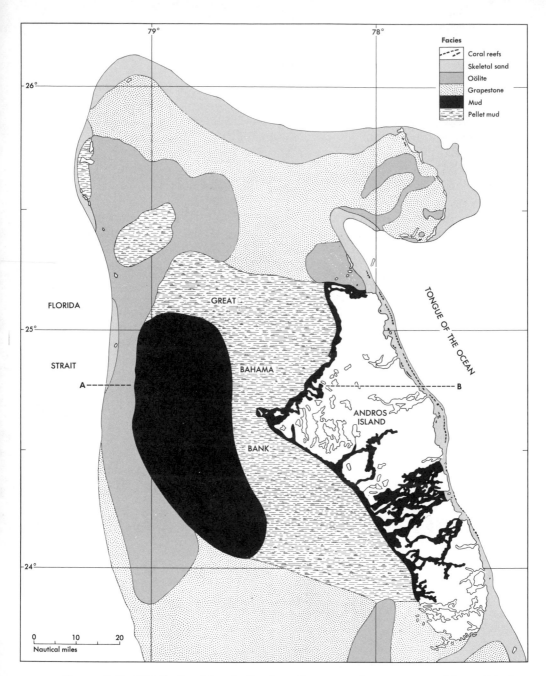

FIGURE 3–8 *Distribution (above) of major sediment or substrate types and (opposite) of bottom-dwelling marine invertebrate communities of the Great Bahama Bank. Note the relatively close coincidence of the distribution of sediment type with the distribution of the bottom communities. On page 48 the cross-section (along line A–B of map above) of the shallowly submerged platform that forms the Bank indicates the importance of salinity, current velocity, and distance from deep, open ocean at the Bank's margin in determining substrate and, by implication, the bottom communities. The* Acropora *community in-*

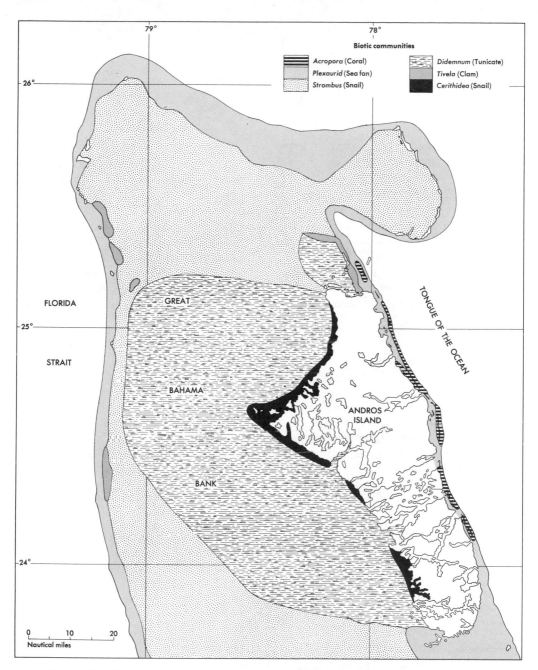

cludes corals, calcareous algae, and calcareous protozoans; the Plexaurid *community includes sea fans, sea whips, snails, scattered corals, sponges, and algae; the* Tivela *community includes a clam, several sea urchins, and marine grasses; the* Strombus *community includes a variety of molluscs and echinoderms; the* Didemnum *community includes a tunicate, green algae, and sponges; the* Cerithidea *community includes a few molluscs, worms, and algae. (After N. D. Newell and others, 1959, and E. G. Purdy, 1964.)*

Organisms and environments

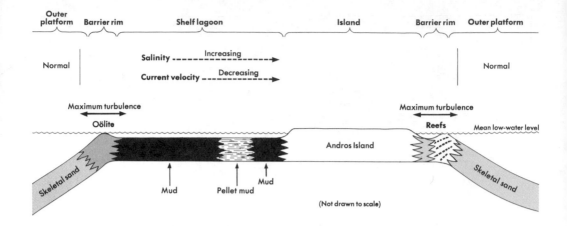

This example suggests the importance of careful observation and analysis of the rock matrix in which fossils are found—as well as of the fossils themselves —in understanding the major physical-chemical factors operating in an ancient environment.

Some Important Environmental Factors: Biological

Organisms are limited in their abundance and distribution by various biological influences or agents as well as by the physical-chemical factors that we have just considered. These biological factors can be broadly summarized by the feeding relationships found among organisms. These relationships include the food-gathering mechanisms of individual organisms, predator-prey interactions, symbiosis, and food chains and webs. Our discussion here will mainly consider animals since plants make their own food through photosynthesis. The major *feeding types* among animals are:

1. **Herbivores,** which consume vegetal matter. These include such diverse animals as cattle, which graze on grasses; snails that browse on the algal films on rocky intertidal surfaces; various insects that feed on the nectar of many flowering plants; and diverse groups of rodents, which eat grain and nuts.

2. **Carnivores,** which prey on living animals—either herbivores or other carnivores. Besides the obvious carnivorous types such as lions, sharks, and eagles, this category contains such different creatures as starfish, the oyster-drill snail, and the voracious shrew. Also in this group are the carrion-feeders, which, like the hyena and vulture, eat the flesh of recently killed animals.

3. **Deposit feeders,** which feed on the organic debris and detritus accumulating on or within the substrate. This group is subdivided into the *selective* deposit feeders, which discriminate between certain kinds of organic matter and avoid as much as possible ingesting the surrounding sediment, and the *nonselective* deposit feeders, which consume the loose substrate for its included organic matter, containing large numbers of unicellular algae, bacteria, and

other microorganisms along with organic molecules and tidbits of decomposing animal and plant tissue. Nonselective deposit feeders void as feces the great bulk of this ingested material, digesting but a small fraction of the total volume. Selective deposit feeders include the many scavenging animals, such as certain snails, catfish, and most crabs. Among nonselective deposit feeders are various terrestrial and aquatic annelid worms, sea-cucumbers (holothurian echinoderms), and some clams.

4. **Suspension feeders,** which, by various processes, strain or filter out of water the suspended organic matter or microscopic organisms. Suspension feeders are as diverse as the baleen whale, which sieves out through the long, fibrous filaments growing from its upper jaw the many thousands of small swimming arthropods living in the sea; many species of clams that bathe their gills with a flow of water from which suspended organic matter is gathered up and brought to the mouth by strands of mucus; and the sponges, which pump water through their elaborate canal system lined with special cells for captureing and ingesting fine-grained suspended food particles.

5. **Omnivores,** which are varied feeders employing two or more of the above feeding types, depending on the availability of a particular food. Man, as well as most other primates, is included within this category.

Obviously, the distribution of these various feeding types will depend on the distribution of the requisite food source. Thus, sedentary suspension feeders prefer water that is sufficiently agitated to ensure that organic matter remains in suspension and is continually transported within their feeding range. The distribution of carnivores parallels the distribution of their prey. Of course, carnivores may be rather far-ranging, moving into different, specific areas to find their prey. For example, a mountain lion will descend into several different valleys to seize grazing cattle and other herbivores. Still, the over-all distribution of mountain lions coincides with that of their prey. Moreover, the abundance of the prey may also control the relative abundance of the predator. Declines in the prey population, either because of too much predation or from some other extrinsic factor, may cause a corresponding decline in the numbers of predators (Fig. 3–9).

Besides predator-prey relationships, there are other feeding associations between different species. In fact, some of these feeding associations may be so close and constant that they are treated as a separate biological phenomenon— *symbiosis,* which literally means "life together." Symbiotic relationships can be divided into three types: those that provide mutual benefit to the participating species ("mutualism"); those that benefit just one participant with little or no advantage gained by the other species ("commensalism"); and those that benefit one species to the detriment of the other ("parasitism").

An interesting example of mutualism is provided by certain green and brown unicellular algae that live within the tissues or protoplasm of some coelenterates and molluscs. In the giant clam *Tridacna* of the tropical Indo-

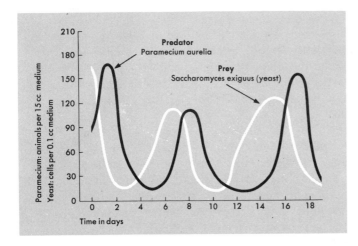

FIGURE 3–9 *Predator-prey relationship between the ciliated protozoan* Paramecium aurelia *and the yeast* Saccharomyces exiguus. *Notice how the abundance of each is related to the abundance of the other. Predation by the protozoan causes a decline in the yeast which in turn limits the number of protozoans. As the protozoans become fewer in number the yeast increases in abundance owing to reduced predation. But soon after the yeast increases its abundance the preying protozoans also increase in number because of the expanded food source. So long as the protozoans do not completely "overeat" their prey, the cycle will continue indefinitely. (From Simpson and Beck, after Gauss, 1965.)*

Pacific and in many shallow-water stony corals, these algae abound in profusion within the soft parts of the host. The algae gain protection as well as additional carbon dioxide and nutrients for photosynthesis. The host gains from the association, too, because it receives photosynthetic oxygen and its nitrogenous wastes are removed. The animal may even obtain some organic nutrients from the algae, although this has not yet been conclusively demonstrated.

Commensalism, which etymologically means "feeding together," is illustrated by the marine sand worm, *Clymenella torquata,* commonly found along the eastern shores of North America. This nonselective deposit-feeding polychaete worm builds a tube of mucus-cemented sand grains that projects from the sediment surface downward for several centimeters into the sediment. Living head down in this tube, the worm feeds on the sediment at the lower end of the tube and it discharges feces from the upper end. Commonly, a small, suspension-feeding clam, *Aligena elevata,* attaches itself to the lower opening of the tube. The clam is taking advantage of the improved water circulation generated by the feeding activities of the worm. The water contained in the sediments, even at this shallow depth of a few centimeters, is not usually well circulated and consequently the content of oxygen and nutrients is diminished. By attaching itself to the worm tube, the clam has more oxygen and food available to it than it might otherwise have.

Parasitism is seen in the adult lamprey, which fixes itself directly to various marine and fresh-water fish by means of a sucker disk around the mouth. The lamprey is attached to the host continuously, feeding on its soft tissue until the host dies. The lamprey seeks out a new victim and the process is repeated.

The dependence of animals ("consumers") on plants ("producers") is evidenced very simply in the case of herbivores such as cattle which feed directly on grass. But since not all animals are herbivores, there is often a sequence, or chain, of feeding relationships leading from the producers to ultimate consumer. Each link in this *food chain* supports the next, starting with the photosynthesizing plants, passing through a number of intermediary animals, and finally ending with a particular animal (Fig. 3–10).

FIGURE 3–10 *Food chain of the herring, showing the various links from the producer organisms of several different kinds of phytoplankton (small floating plants) through a link of zooplankton (small floating animals) composed of different marine invertebrates, to the final link of the adult herring. Notice that the juvenile and smaller herring feed on the smaller zooplankton. (From Hedgpeth, 1957.)*

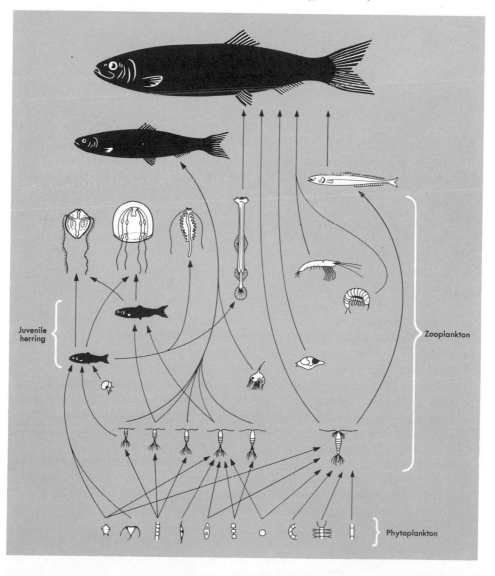

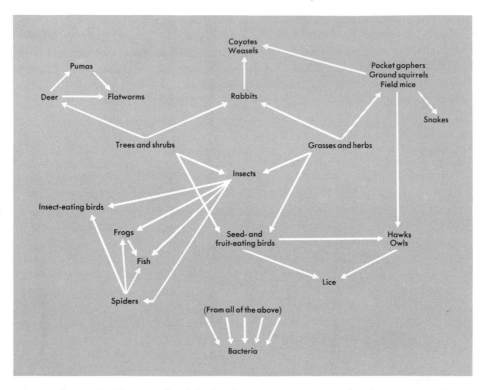

FIGURE 3–11 *Highly generalized food web in an American woodland community, show-ing the flow of energy through various feeding relationships. The actual interactions among these organisms and others not included in this simplified diagram are far more complex. This complexity is further compounded when we consider the various physical-chemical factors that influence the abundance and distribution of these different species. (From Simpson and Beck, 1965.)*

In any given environment there may be many such food chains, each of which may include some links in the others. The result is a complex interaction —*food web*—among the animals and plants living in this environment (Fig. 3–11).

It should be clear, therefore, that in virtually all natural environments the abundance and distribution of any particular species of organisms are controlled or influenced by a variety of physical, chemical, and biological factors. As the number of species increases these various interactions among organisms become compounded. Because of these complex interrelationships among organisms with one another and with their inorganic surroundings, it is extremely difficult to isolate any single factor as the most important one. Yet, despite this complex-ity, it is sometimes possible to define some one ecologic factor as being particu-larly critical, or limiting, so that most of the others either hinge upon it or are by comparison somewhat less important. This concept of the *limiting factor* is often referred to as "Liebig's Rule," after the nineteenth-century biologist Justus Liebig who first formulated it.

An example of a limiting factor is the amount of available nitrates and phosphates in an aquatic environment, where the total abundance of the plants, and by implication the animals, depends quite closely on these nutrient substances. Should the availability of these nutrients fall below some minimum level, the plants and animals would perhaps be severely affected. It should be emphasized, however, that recognition of a limiting factor in an environment does not imply that other ecologic factors can be ignored altogether.

Assemblages of Organisms

We have seen how the distribution and abundance of *individual* organisms are controlled by the physical, chemical, and biologic conditions of the surrounding local environment. It follows, therefore, that *aggregations* of organisms will have similar ecologic control. Consequently, the composition of a given assemblage or aggregation of animals and plants at a particular locality will reflect the environmental conditions existing at that place.

In some instances the aggregations of different species in nature may be the result of overlapping ecologic tolerances of the individual component species (Fig. 3–12). In other cases these aggregations may represent a complicated system of energy transfer operating through several different feeding levels (Fig. 3–11). Assemblages of organisms, therefore, although definitely influenced by the local environment, are either "statistical associations" or "integrated

FIGURE 3–12 *Schematic diagram to illustrate how two different assemblages of organisms may reflect different overlapping ranges of ecologic tolerances of component species. In both cases the association of the individual species is the result of mutual ecologies with a minimum of species interaction.*

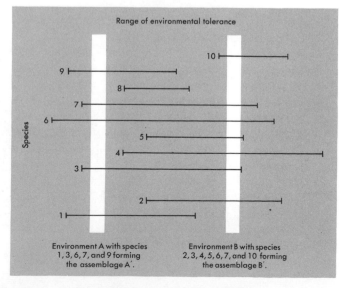

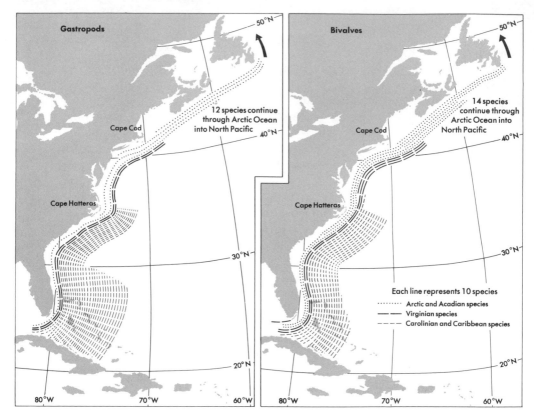

Gastropods

12 species continue
through Arctic Ocean
into North Pacific

Cape Cod

Cape Hatteras

Bivalves

14 species
continue through
Arctic Ocean into
North Pacific

Cape Cod

Cape Hatteras

Each line represents 10 species
············· Arctic and Acadian species
—— Virginian species
— — Carolinian and Caribbean species

FIGURE 3–13 (Above) Diversity gradients of
molluscs along eastern North America. (At
left) The numbers of species decrease with
decrease in ocean temperature. (Below) A
similar diversity gradient in the numbers of
genera of a family of Permian brachiopods.
(From Fischer, 1960, and Stehli and Helsley,
1963.)

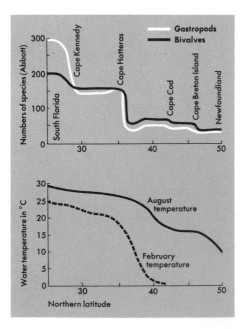

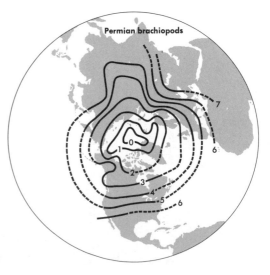

Permian brachiopods

communities." The actual degree to which assemblages of organisms are either one or the other is a source of debate among ecologists. The point to remember, however, is that whatever the degree of species interaction, the composition of the assemblage is environmentally controlled.

Besides local differences in the composition of organic assemblages owing to local environmental influences, there are also some interesting regional or geographic variations in such assemblages. For example, the *diversity* of animal and plant species increases away from the Earth's polar regions toward the equator. Thus, in the boreal climates, forests are usually composed of only a few species of trees such as pines and birches. In contrast, tropical forests may have many hundreds of different plant species. Similar latitudinal diversity gradients have been observed for many other organisms, including birds, snakes, various marine invertebrates, and even for some fossil assemblages (Fig. 3–13). The *numbers of individuals* of a particular species, however, increases with decreases in diversity of species toward the poles. This gradient may merely reflect an increase in food resources and living space with the decline in number of species.

Interpreting these diversity gradients is not as easy as observing them. One suggestion is simply that numbers of species decrease because few organisms are able to adapt to the rigors of boreal climates. An alternate hypothesis states that the tropical habitats have been fairly stable and so have had a long time for developing diversity. The polar regions, on the other hand, have had drastically fluctuating environmental conditions, as during the glacial and interglacial epochs of the Pleistocene Epoch. Consequently, these polar regions have not had an equal length of time to develop a rich, flourishing variety of life.

Another kind of diversity gradient, but on a smaller scale, is associated with systematic ecologic variations. For example, salinities in the Baltic Sea decrease markedly from the North Sea toward the Gulfs of Finland and Bothnia. Paralleling this salinity decrease is a sharp decline in the number of species of marine animals (Fig. 3–14). A similar decrease in species diversity has been observed among fossils recovered from late Pleistocene deposits from the Lake Champlain and St. Lawrence River valleys (Fig. 3–15 and Table 3–1).

Biogeography

It is undoubtedly apparent to anyone who has traveled extensively that animals and plants have particular geographic distributions. Quite obviously, these patterns in biogeography are related to the ecological conditions that vary from place to place throughout the Earth. We have already discussed, for example, how the composition of marine invertebrate communities changes with latitude. Thus, coral reef assemblages are restricted to the warm, shallow,

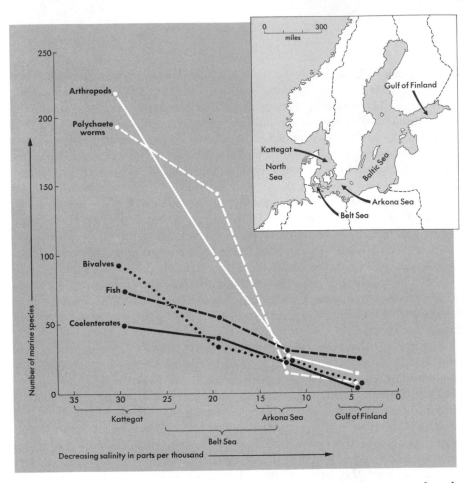

FIGURE 3–14 *There is a sharp decrease in the number of marine species as we go from the North Sea through the Kattegat into the Baltic Sea and the Gulf of Finland. This decrease in diversity of marine species parallels a rapid decline in water salinity. (Data from S. G. Segerstrale, 1957.)*

well-lit waters of the low latitudes. Fig. 3–16 (p. 58) illustrates major plant communities; these are differentiated mainly by regional ecological variations. Therefore, besides defining local assemblages of organisms in terms of the local environment, we can also recognize characteristic assemblages on still a broader geographic scale. In a sense, then, biogeographic studies are synecologic in essence, but on a larger spatial scale.

As we shall see, however, biogeography is a result not only of ecology but also of other important factors, such as history and barriers to dispersal. For example, some organisms may be prevented from occupying a suitable environment because they cannot breach barriers to their dispersal. Such barriers for land dwellers include mountain ranges, deserts, and large bodies of water. Marine organisms, on the other hand, may be prevented from invading a suitable

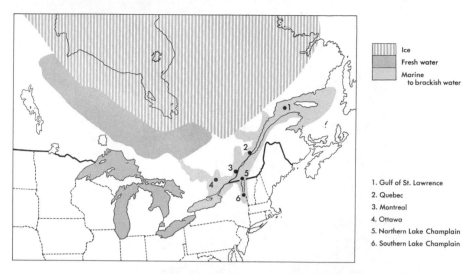

FIGURE 3–15 *The maximum extent of the Champlain Sea during Late Pleistocene time. Salinities of the Champlain Sea decreased away from the open ocean. Paralleling this salinity decrease was a decline in the number of marine species (see Table 3–1). (From A. La Roque, 1949.)*

environment because of intervening land masses or widespread oceanic deeps. Thus, the unique character of the flora and fauna of Australia is a result of its being isolated from the rest of the continents for millions of years. Similarly, the shallow-water marine invertebrates in the Indo Pacific region differ considerably from those on the western shores of the Americas because of the great expanse of deep water of the eastern Pacific Ocean separating these two areas. Few bottom-dwelling marine invertebrates have sufficiently long-lived floating larval stages to survive the trip necessary to go from the western to the eastern

Table 3–1

Decrease in Diversity of Marine Species
Going Away from Open Ocean

Number of Marine Species	Localities					
	1	2	3	4	5	6
Foraminifera	18	13	15	2	—	—
Sponges	1	—	2	1	—	—
Echinoderms	2	1	5	1	—	—
Bryozoans	26	5	2	1	—	—
Brachiopods	3	1	1	—	—	—
Clams	28	12	18	10	5	1
Snails	41	12	40	7	—	—
Annelid worms	11	1	2	2	—	—
Crustaceans	4	3	2	1	1	—
Total	134	48	87	25	6	1

From W. Goldring, 1922, N.Y. State Mus. and Sci. Bull. 239, p. 164.

Organisms and environments

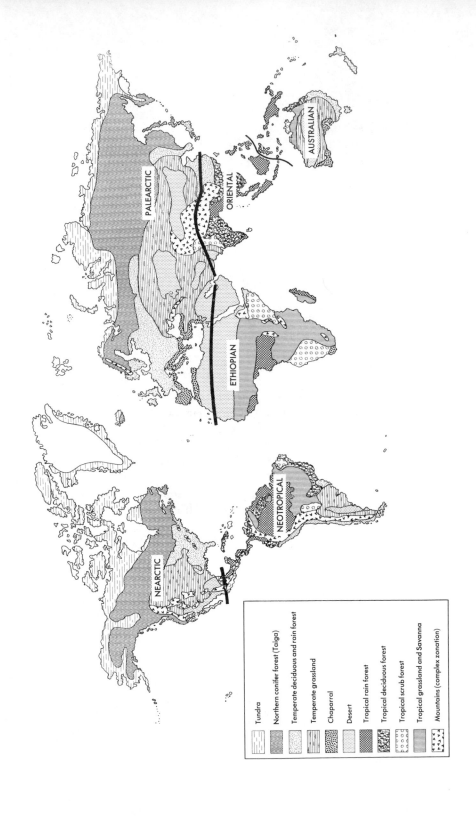

PALEARCTIC

ORIENTAL

ETHIOPIAN

AUSTRALIAN

NEARCTIC

NEOTROPICAL

Tundra

Northern conifer forest (Taiga)

Temperate deciduous and rain forest

Temperate grassland

Chaparral

Desert

Tropical rain forest

Tropical deciduous forest

Tropical scrub forest

Tropical grassland and Savanna

Mountains (complex zonation)

FIGURE 3–16 *(Opposite page) Distribution of the major plant communities around the world. The character of these communities—as well as that of the animal communities associated with them—depends not only on the environment of the region, but also on its history. These ecological and historical factors result in the differentiation of six major biogeographic regions; Nearctic, Palearctic, Neotropical, Ethiopian, Oriental, and Australian. (From E. D. Hanson, 1964.)*

Pacific. The lack of abundant islands in the eastern Pacific also inhibits island-hopping across the several thousands of kilometers of water.

Biogeography has another important element, besides ecological and dispersal factors—the historical element. Because the geography of the Earth has changed through time, the distribution of organisms has correspondingly changed. Consequently, the flora and fauna of any one place at a particular time are a result of the existence of earlier organisms in that place (historical factor) as well as the environmental conditions found there (ecological factor).

A revealing example of historical factors in determining biogeographic patterns is offered by the South American mammals. In the early Tertiary Period, South America was connected to North America by the Isthmus of Panama and was populated by a variety of primitive mammals. Shortly thereafter, the Isthmus of Panama was submerged under the sea and South America became isolated from the rest of the world. During this period of isolation the ancestral mammals of early Tertiary time evolved into a number of different groups, including various types of terrestrial herbivores and carnivores. The isolation of South America was not complete, however, because a few animals were introduced around the middle of the Tertiary Period. These forms included certain species of New World monkeys and rodents which, apparently, island-hopped from North and Central America to the South American mainland.

Toward the close of the Tertiary Period, the isolation of South America, which had lasted some 40 to 50 million years, was ended by the re-establishment of the land connection between North and South America. Mammals of both regions then began to spread northward and southward, migrating from one continent to the other. Not all species migrated, however; nor were those that did equally successful in the new continent, whether North or South. The North American mammals, though, were able before long to occupy many of the niches previously occupied by the indigenous South American species, leading to the replacement and extinction of many of the South American mammals (Table 3–2).

The present-day composition of the South American mammalian faunas is, therefore, a result not only of the particular environmental conditions found there, but also of the evolutionary history of the area during the Tertiary Period. Thus, the South American mammalian fauna includes groups that were initially established there in the early Tertiary Period, such as the armadillo, tree sloth, and anteater; monkeys and rodents that were somehow able to cross the water

Organisms and environments

barrier separating the two continents during the middle Tertiary Period; and late Tertiary arrivals from North America, such as field mice and various cats, as well as groups such as llamas and tapirs that have since become extinct in North America.

Table 3–2

Characteristic Families of Land Mammals In South and North America*

Epoch	South America			North America		
	Total	Indigenous	North American	Total	Indigenous	South American
Recent	30	16	14	23	20	3
Pleistocene	36	23	13	34	26	8
Pliocene	25	24	1	27	26	1
Miocene	23	23	0	27	27	0

From G. G. Simpson, 1953, *Evolution and Geography,* Ore. State System of Higher Educ., Condon Lectures, p. 27.

*The table shows the total isolation of North and South America during the middle Tertiary Period (Miocene Epoch), with each continent having its own characteristic mammalian families. During the Pliocene Epoch the land connection between the two continents was re-established and there was considerable faunal interchange. North American mammals, however, replaced many of the indigenous South American forms. Consequently, about one-half of the present-day mammalian families in South America have North American origins, whereas less than 10 per cent of the North American families have a South American ancestry.

4

Geochemical

environmental evidence

So far we have considered the interactions of organisms with their environment as well as the relationships of sediments to the environment. We now want to review how geochemical evidence can be used for environmental interpretation. The question we will be posing is this: What kind of chemical evidence can be obtained from sedimentary rocks and their included fossils that will tell us something significant about the environment in which they formed?

Although geochemical environmental evidence has received considerable attention recently, there is not as yet any generally reliable body of geochemical data or a set of procedures that gives unequivocal environmental evidence. The reason is that, although a given chemical system may be in equilibrium with a particular sedimentary environment (thereby perhaps uniquely characterizing that environment), postdepositional changes in environment establish new chemical equilibria. Thus, even small changes in temperature, pressure, and water content and mobility in the interstices of a sediment, can result in significant changes in the chemical attributes of a sediment or rock. The new, postdepositional

chemistry will mask or replace the original chemistry that might have characterized the depositional environment. Indeed, the chemical attributes of a sediment that might identify a specific environment are usually far more susceptible to postdepositional alteration or obliteration than are fossils or the physical features of a sediment. Of course, chemical—and, by extension, mineralogic—evidence will often record and explain the postdepositional history of a rock. Yet, despite these inherent difficulties, sedimentary geochemistry has considerable promise in identifying and interpreting original environments.

Geochemical Evidence from Calcareous Skeletons

Heinz Lowenstam of the California Institute of Technology has shown that many calcium-carbonate-secreting marine algae and invertebrates form different mineral varieties of $CaCO_3$—namely, aragonite or calcite—depending on local water temperature (Fig. 4-1). Some invertebrates, such as various clams, snails, bryozoans, and serpulid worms, will secrete $CaCO_3$ both as calcite and aragonite, but will vary the ratio with changing water temperatures. Thus, in cold water certain forms will have high calcite-to-aragonite ratios, whereas in warmer waters the same species will have lower calcite-to-aragonite ratios.

Other groups of organisms, such as stony reef-building corals, which secrete only aragonite, will have many warm-water species but only a few cold-water varieties. Certain aragonite-secreting red algae and alcyonarians (a group of colonial coelenterates) will occur only in warm water, while closely related, calcite-secreting species will be found in both cold and warm water.

Although the chemical dynamics are not fully understood, it is clear that the secretion of $CaCO_3$ in the form of the mineral aragonite is definitely favored in warmer marine waters. Therefore, an increase in the abundance of ara-

FIGURE 4-1 *Schematic diagram showing three major kinds of calcite-aragonite variations among marine calcareous algae and invertebrates. Group 1 includes organisms that secrete only aragonite, but that have more species in warmer waters than in colder waters (for example, reef-building corals). Group 2 includes organisms that secrete both calcite and aragonite, although the aragonite species are restricted to warm waters (for instance, certain algae and alcyonarian coelenterates). Group 3 includes organisms that individually secrete both calcite and aragonite in the same shell but with the ratio changing with temperature (such as many clams, snails, bryozoans, and serpulid worms.) (From H. Lowenstam, 1954.)*

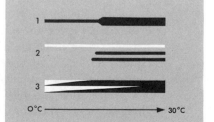

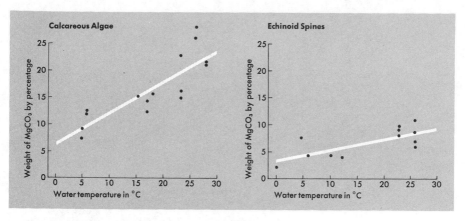

FIGURE 4–2 *The increase of magnesium in the mineral calcite with increasing water temperature, in calcareous algae and echinoid spines. Magnesium substitutes for calcium in the crystal lattice of the mineral calcite, $CaCO_3$, forming $MgCO_3$. Note that the amount of magnesium substitution is greater for the calcareous algae than it is for echinoids over the same temperature range. (From K. Chave, 1954.)*

gonitic fossils in an ancient rock would, other things being equal, indicate increasing water temperatures of the water in which those organisms once lived. But "other things" are not usually equal. For example, the mineral aragonite, unlike calcite, is not stable, and with time it spontaneously reverts back to the more stable crystal structure of calcite. Rocks generally older than Cretaceous age, therefore, contain less and less aragonitic fossils. Paleozoic rocks, with one single exception, lack aragonitic fossils altogether. Consequently, this temperature relationship between mineral and water is only potentially useful for Cretaceous and younger fossils.

Keith Chave of the University of Hawaii has investigated a similar chemical variation with changing water temperature. He discovered that, with increasing water temperatures, many marine algae and invertebrates that secrete calcite will vary the amount of magnesium, which substitutes for some of the calcium in the calcite mineral structure. Within a given group of organisms, higher magnesium values are found in the warmer-water part of the organism's range than in the colder-water part (Fig. 4–2). But here too this relationship has been hard to establish in ancient fossils, for high-magnesium calcite is very unstable and after burial quickly converts to low-magnesium calcites as the magnesium is leached out by water in the surrounding sediments.

Evidence from Isotopes

Organically secreted calcium carbonate may contain other clues, besides mineralogic ones, regarding the temperature of the surrounding water. The element oxygen—which is, of course, a constituent of calcium carbonate ($CaCO_3$)—has several different isotopes. These isotopes, which are chemically identical

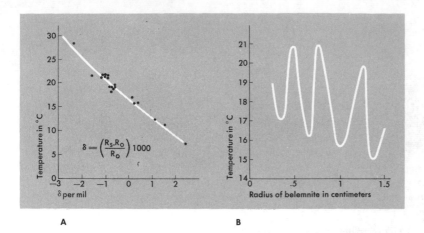

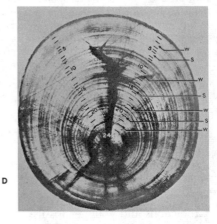

A $\quad\quad\quad\quad\quad\quad\quad\quad\quad\quad$ B

$$\delta = \left(\frac{R_s\text{-}R_o}{R_o}\right) 1000$$

C

D

FIGURE 4–3 (A) Decrease in the relative abundance of O^{18} with increasing water temperature. $R_s = O^{18}/O^{16}$ of sample; $R_o = O^{18}/O^{16}$ of standard. (B) Temperature variations observed in Mesozoic belemnite (an extinct cephalopod) based on oxygen isotopes. Periodic high and low temperatures reflect summer warming and winter cooling of the sea during the life of the belemnite. There is also a suggestion that the surrounding average water temperature declined as the animals aged, indicating that the belemnite migrated to cooler waters or lived at greater water depths. (C) The calcareous skeleton of a belemnite of Jurassic age from North Dakota. (D) Cross-section of belemnite, showing darker and lighter winter (W) and summer (S) growth layers. Annual layers are numbered. (From M. Kay and E. Colbert, 1965.)

varieties of oxygen though they have different atomic masses, are designated as oxygen 16, 17, and 18, depending on the total number of protons and neutrons in the nucleus.

Although these isotopes have relative abundances of 99.76, 0.04, and 0.20, (O^{16}, O^{17}, and O^{18}) respectively, these relative abundances vary slightly but perceptibly with temperature. Thus, the O^{18}/O^{16} ratio of ocean water varies indirectly with increases in water temperature. For example, sea water at $10°C$ will have a relatively higher O^{18}/O^{16} isotopic ratio than sea water at $30°C$. Having once established the relationship of O^{18}/O^{16} ratios with sea water tem-

Geochemical environmental evidence

peratures of present-day oceans, we can calculate the temperatures of ancient seas by measuring the O^{18}/O^{16} ratio in the calcareous portion of fossil shells (Fig. 4–3). Several such studies have shown that Jurassic and Cretaceous seas, although somewhat warmer than those of today, had oscillating temperatures and that the Late Mesozoic tropical and subtropical zones were wider than today's and had more equable climates. As will be shown in the last chapter, oxygen-isotope ratios, as determined from the shells of minute calcareous organisms that float in the sea, have provided evidence for successive periods of cooling and warming of the North Atlantic during the glacial and interglacial epochs of the Pleistocene.

Other isotopes can provide additional information about ancient environments. For example, the C^{13}/C^{12} ratio found in land plants varies considerably from that of marine plants. Shells, too, secreted in sea water have different ratios of carbon isotopes from those of shells secreted in fresh water (Fig. 4–4). Determination of C^{13}/C^{12} ratios in ancient rocks and fossils, therefore, should suggest the nature of the original environment in which these sediments were laid and fossils formed, if this wasn't already apparent in the first place. For example, geochemists have measured the carbon-isotope ratios of selected marine and fresh-water sediments and compared these results with the ratios obtained from a number of Devonian crude oils. Surprisingly enough, even though the crude oils were obtained from marine rocks, they have carbon-isotope ratios that suggest that their organic matter (which presumably formed the oils) was terrestrial or fresh-water in origin. The geochemists concluded that, if their results were accurately obtained and their reasoning correct, pe-

FIGURE 4–4 *Deviation of C^{13}/C^{12} in natural systems. $R_s = C^{13}/C^{12}$ sample; $R_o = C^{13}/C^{12}$ standard. Using these data for reference it is possible to determine the origin of a sample of ancient carbon-containing material as being from one of these systems. For example, carbon-isotope analysis of 128 Devonian crude oils gives a δ of −21 o/oo to −32 o/oo, with an average of 27.5. This would suggest that the oils are derived from organic material formed either in the atmosphere or in fresh waters on land. (From H. Craig, 1953; and W. R. Eckelmann, W. S. Broecker, D. W. Whitlock, and J. R. Allsup, 1962.)*

$$\delta = \left(\frac{R_s - R_o}{R_o}\right) 1000$$

troleum most likely develops in marine sedimentary basins marginal to land areas that provide sizeable quantities of organic detritus.

Carbon isotopes have also been used to determine whether marine rocks have been subjected to fresh-water alteration. For, if marine sediments with high initial values for the C^{13}/C^{12} ratio are lithified and uplifted above sea level where they can be permeated by fresh water, then their carbon isotope-ratio will be enriched in C^{12}, resulting in a relatively lower C^{13}/C^{12} ratio. Geologists have used this relationship to demonstrate that Pleistocene marine limestones ("marine" because of their abundant marine fossils) in the Bahamas and Bermuda have been exposed to air and permeated by fresh ground water. The carbon-isotope variations consequently suggest that these limestones may have been geochemically altered in other significant ways, too. Such alterations might seriously throw into question any other geochemical data obtained from these rocks regarding the nature of the original marine environment of deposition.

Trace Elements, Organic Compounds and Environments

Clay minerals have a characteristic property of taking up certain elements such as boron, gallium, and rubidium into their crystal lattices. Inasmuch as the relative abundance of these elements varies from fresh water to sea water, a number of investigators have used the proportions of trace elements within clay-rich rocks, such as shales and mudstones, to determine their environment of deposition (Fig. 4–5).

The element boron also seems to have an affinity for the clay mineral illite, and some geologists have tried to use boron alone as an indicator of paleosalinity. Since the boron content of natural waters is proportional to their salinity, it has been reasoned that fresh-water, illite-rich sediments or rocks will have lower boron values than those deposited in marine environments.

One of the more recent and exciting developments in geochemistry that relates to organisms and environments has been the extraction and identification of organic molecules of presumed biologic origin from very old rocks. Although strictly speaking these organic compounds do not define or delimit particular environments, their very presence in ancient rocks that are not otherwise fossiliferous does indicate a former biologic environment.

These organic compounds, which have been colloquially termed "chemical fossils," include *phytane* and *pristane*, which are degradation products of chlorophyll, a green pigment in plants. Although the presence of pristane and phytane does not absolutely prove the former existence of chlorophyll (and hence, by extension, of plant life) in the original depositional environment of these rocks, geochemists consider it extremely unlikely that these particular

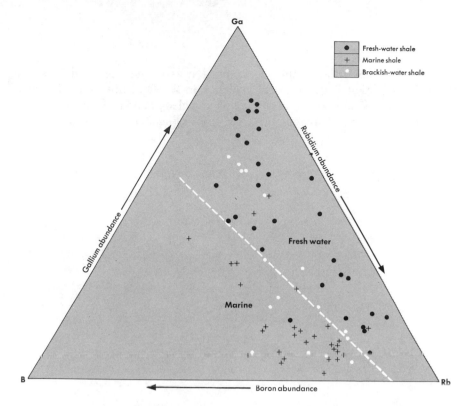

FIGURE 4–5 *Relative proportions of the trace elements boron, gallium, and rubidium in Pennsylvanian shales of the Appalachians. The shales were initially defined as fresh-water, brackish-water, or marine on the basis of their associated fossils. As the graph shows, the fresh-water shales are relatively richer in gallium, whereas the marine shales are richer in boron and rubidium. (From E. T. Degens, E. G. Williams, and M. L. Keith, 1957.)*

compounds could be inorganically produced. If this assumption is correct, then the recent discovery of pristane and phytane in rocks that are some 3 billion years old extends the antiquity of life back at least another 1 billion years. This would mean, therefore, that environments suitable for photosynthetic organisms have been in existence for a very, very long time. Moreover, these green plants would have provided a food source for other organisms ("animals"). Fossil animals, however, have not yet been identified from these rocks. One wonders why. Had animals not yet evolved, or were they present as small, unfossilizable forms? Perhaps these fossils are there but have not yet been recognized as such. In any case, these organic compounds indicate the presence of biologic environments in rocks that had been considered to antedate the first appearance of life.

In summary, we can see that significant geochemical data regarding the nature of ancient environments are contained in the geologic record of rocks and fossils. So far, however, most of the data raise at least as many questions, uncertainties, and problems as they solve. The chief reasons for these difficul-

Geochemical environmental evidence

ties are the complexities of the geochemical systems themselves, the relative ease of losing original geochemical evidence in the later history of the rocks and fossils, and finally, the general lack of knowledge about the geochemistry of recent sedimentary environments and their associated organisms.

5

Environmental analysis

In previous chapters we have reviewed some of the relationships among organisms, sediments, and the local depositional environment. But knowledge of these relationships alone will not enable us to reconstruct an ancient environment. We also need to develop some insight into just how we go about collecting and examining fossils and rocks so that we can arrive at meaningful conclusions and significant interpretations about ancient life and its environmental setting.

In this chapter, therefore, we will consider some of the assumptions, procedures, viewpoints, and limitations that form the background for paleoecological investigation and interpretation—to ask, in other words, in what sort of general intellectual framework are paleoecological studies pursued? In earlier chapters we have discussed paleoecological tactics; here, we will discuss paleoecological strategy.

Uniformitarianism

Up until about the beginning of the nineteenth century it was generally believed that fossils within rocks were sports of nature, works of the devil, or the remains of animals and plants buried by the Biblical Deluge. Among the more educated, however, fossils were indeed recognized as the petrified remains of once-living organisms. But it was commonly assumed that these organisms had been killed by successive and sudden geological "revolutions" during which the seas inundated the continents, oceans dried up, and the Earth's surface underwent great upheavals. Such beliefs were summed up by paleontologist Georges Cuvier (1769–1832) in his theory of *Catastrophism*.

At about the same time several Scottish geologists, including James Hutton, James Hall, and John Playfair, were in the process of formulating an opposite point of view, namely that the Earth undergoes constant change that takes place inexorably over great spans of time. The death knell of the catastrophic school of geology was sounded by the publication in the 1830's of the *Principles of Geology* by Charles Lyell, another British geologist. Lyell's particular contribution to the development of the young science of geology is best summed up in the subtitle to this work: ". . . an attempt to explain the former changes of the Earth's surface by reference to causes now in operation." Thus Catastrophism, which emphasized sudden and violent changes in Earth processes during Earth history, was replaced by *Uniformitarianism*, which emphasized slow, regular changes by processes similar to those observable today.

Since Lyell's time, geologists have realized the value of studying present-day Earth processes in order to understand ancient geologic phenomena. As Lyell himself expressed it, "the present is a key to the past." Although there are limitations to such an approach (for there may be some past geologic events that were truly unique and have no modern counterpart), paleoecology has depended strongly on this uniformitarian viewpoint. In fact, most paleoecologic studies rely heavily on modern environmental processes and phenomena to support inferences about the geologic past (Table 5–1). This, then, is one of the first axioms of paleoecology, that until proven otherwise ancient geologic environments have modern analogues. We can conveniently refer to this assumption as the *rule of analogy*.

Necessity of a Pluralistic Approach

Another important concept in paleoecological analysis is the realization that multiple lines of independent evidence must be sought and developed before any final conclusions are drawn about the nature of an ancient depositional environment. Because of the inductive and inferential nature of paleoecology, we can never definitely prove the truth of our assertions about past environments and communities of organisms. The only criteria we have for believing the validity of our paleoecological interpretations are, first, the internal consistency of multiple sets of independent data which lead to the

Table 5–1

An Ancient Environment and Its Modern Analogue*

Manlius Formation (Devonian, New York State)	Recent Analogue (South Florida and Great Bahama Bank)
1. Algal-laminated structures	Intertidal and shallow subtidal level in Florida Keys and Andros Island, Bahamas
2. Laminated, calcareous mud with pellets; often air-dried and mudcracked.	Supratidal areas in Florida Keys, islands in Florida Bay, and Andros Island
3. Interbedded shell sands and calcareous muds: some dried mud-pebble conglomerates	Intertidal zone in Florida Keys and islands in Florida Bay
4. Scattered calcareous oölite sand	Just below intertidal zone, margins of Great Bahama Bank

After L. Laporte, 1967, *Am. Assoc. Petroleum Geologist's Bull.*, v. 51, p. 90.

° Various rock and fossil characteristics of a 400-million-year-old limestone deposit of New York State and the depositional environments of their present-day counterparts as found in South Florida and the Great Bahama Bank. This comparison of the ancient and Recent suggests strongly that the Manlius was deposited in warm seas of very little depth, that is, at (or slightly above or below) mean sea level.

same final conclusions, and, second, the geologic and biologic sense our interpretation makes when compared to present-day environments and organisms.

Although it is the essence of the scientific method to formulate simple hypotheses to explain natural phenomena, we cannot simply perform a single set of critical observations and expect that all will become clear. However astute the observer and obvious the phenomena, geologic situations unfortunately have a tendency to be the result of many interacting variables, many of which are no longer directly recordable.

The Problem of Reduction

As we have just noted, sedimentary environments are inherently complex and multivariable. Hence, attempts to explain the abundance and distribution of organisms within a given environment by reference to any one single factor may be a serious oversimplification and misleading. For example, until recently some paleoecologists have attributed virtually all variation in the composition of

Environmental analysis

fossil marine assemblages of similar age to differences in the original water depth of the ancient environment. Thus, whenever different fossils were encountered in a stratigraphic sequence of essentially the same age in a particular area, these differences were attributed to local changes in sea level. This explanation, in turn, required multiple and frequent sea-level fluctuations, an hypothesis that, for other paleoecologists, strained credulity.

Although it is true that the water depth of present seas sometimes correlates with changing marine communities, there are other, more direct, ecologic factors that limit marine organisms. For example, water temperature, oxygen and nutrient content, light, and substrate are usually more significant than water depth. The reason that depth variation is sometimes correlated with community variation is that these other, more direct ecologic factors are likely to vary with depth. Thus, substrates tend to be finer-grained and light intensity decreases with increasing depth in the oceans. Yet many organic communities vary dramatically without any significant depth changes; the depth-control theory proves useless in explaining these variations. Such instances are the result of differences in critical ecologic factors that are, in these specific cases, independent of depth variation. For example, the marine communities of the Great Bahama Bank vary from area to area, yet these communities may be found in virtually all water depths that occur on the Bank proper. The ecologic factors determining the Bahamian communities are salinity, water turbulence and circulation, and substrate (see Fig. 3–8).

Once we have interpreted one environmental complex of sediments and organisms in terms of specific ecologic factors, we must resist the temptation to explain all similar complexes in the same way, since different environmental processes and controls may give roughly similar results. For example, red beds, or black shales, or reef limestones, each with characteristic lithologic and biologic attributes, may not necessarily form in always the same set of circumstances. It may well be that eventually paleoecologists will provide relatively simple, unifying theories for many different ancient environments. But until that time, the results of individual paleoecological studies must be cautiously applied to other situations, however similar they may at first appear.

The problem of reduction, therefore, should make us wary of either placing too much value on a single environmental factor in a paleoecological interpretation, or assigning one set of paleoecological conclusions, however valid, to a second paleoecological problem merely because the two share some similarity of fossil composition and rock type.

Value of an Environmental Datum

Initial analysis of ancient environments usually yields data and interpretations that are, at best, of relative value only. Thus, we might conclude that this facies was deposited in more saline and quieter water environment than that facies. But what do "more saline" and "quieter" mean? Even if we cannot express the salinity in parts per thousand or the water velocity in centimeters per second, can we be more definite in our interpretations? We can if we are able to establish within our paleoecologic framework an *environmental datum* to which we can refer our other observations.

If, during the course of an investigation, we find a particular lithologic set of characters, or a fossil organism or assemblage, which has modern counterparts that are well understood, then we can establish from this lithologic or biologic feature an environmental datum. Such a datum might be sun-dried mudcracks, coral reefs, fresh-water calcareous algae, oölites, or molds of salt crystals; any of these would pinpoint somewhere within the total paleoecologic complex the existence of certain environmental conditions. Rocks and fossils associated directly or indirectly with this datum could then be related ecologically to it, if only by the elimination of other, incompatible environmental possibilities. For instance, the finding of abundant and well-preserved reef corals in a limestone would immediately rule out a deep-water origin for the rock because reef corals require shallow, well-lit, agitated, and warm waters. The paleoecology of the other organisms that occur with the corals would thereby also be elucidated.

Lateral vs. Vertical Facies Studies

There are two basic approaches taken in paleoecological studies. Each is equally valid but their goals are somewhat different. The first approach seeks to understand the *areal variations* in local sedimentary environments and fossil assemblages. In this way we concentrate on a thin stratigraphic interval that covers a wide geographic area so that several different lithofacies and biofacies of essentially the same time may be examined. The second approach is to see how environments and organisms have changed through time, either in a particular region or throughout the whole world. Different successive stratigraphic sequences are examined to identify *temporal variations* in organic communities and depositional environments.

Lateral studies result in the definition and delineation of the depositional basin together with its various sedimentary environments and fossil organisms. The paleogeography, that is, the distribution of ancient lands and seas, is also established. In addition, such studies permit the recognition of the ecologic relationships among the sedimentary environments and the associated fossils. For example, studies of the reef complex of the Permian Period of West Texas and southern New Mexico area show how reefs developed on the margins of deep-water basins (Fig. 5–1) and formed a series of environments in front of and behind the reefs (Fig. 5–2). Similar lateral facies studies of ancient reef complexes have been made throughout the stratigraphic record and in many parts of the world.

Vertical facies studies attempt to integrate through time the results of lateral facies studies. Thus, it would be extremely interesting to examine various reef communities that have developed during the 600 million years since the Cambrian Period. Reefs are known from the Cambrian (when the dominant reef-forming organisms were sponge-like archaeocyathids), Ordovician (bryozoans and primitive tabulate corals), Silurian, Devonian, and Carboniferous (tabulate and hydrozoan corals), Permian (calcareous sponges and rooted brachiopods), early Mesozoic (primitive scleractinian corals), late Mesozoic (rudisitid clams), and Cenozoic times (advanced scleractinians). Were these all true reefs in the sense that they were wave-resistant structures growing up to mean sea-level? Did the dominant organisms that built the framework in each case fill the same

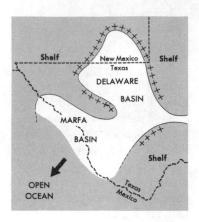

FIGURE 5–1 *Map of western Texas and southern New Mexico during Permian time showing deeper-water basins surrounded on three sides by shallow-water shelf areas. Along the margins of the Delaware Basin, reefs developed. The reef-dwelling organisms (calcareous sponges and algae, rooted and cemented brachiopods, some bryozoans and fusulines) flourished in the shallow, warm, well-lit waters and obtained a constant and rich supply of food and oxygen from the deep waters that surged back and forth at the shelf basin margins. (After N. D. Newell 1957.)*

FIGURE 5–2 *Cross-section of reef-complex deposits of Permian age in western Texas and southern New Mexico. Major zones across reef and reef-associated environments shown as A-G. Black areas represent stagnant marine waters; gray areas represent marine waters of normal oxygen content and salinity. A—Hypersaline waters in lagoon with few, specially adapted organisms. B—Broad shoaling area with crinoids, fusulines, and calcareous algae. C—Reef top with sponges, calcareous algae, rooted and cemented brachiopods, bryozoans, and fusulines. Waters here are of normal salinity and well agitated. D—Stalked brachiopods, delicate bryzoans, and various algae. E—Intermittently stagnant bottom with black, silty muds; a few siliceous sponges, scattered molluscs and echinoids. F—Continuously stagnant bottom with no bottom-dwelling organisms; dead carcasses of pelagic organisms from above may collect here. G—Pelagic fauna rich in micro-organisms (plankton), ammonoid cephalopods, and fish. (After N. D. Newell, 1957.)*

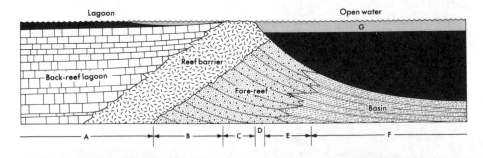

ecologic role in the reef community? Were the community relationships of each of these reef associations essentially the same, with only the taxonomic character of the participants changing? Or were there fundamental ecological differences in these reef communities? Were the reefs that developed well within the continental interior (as in the Silurian reefs of Illinois and Indiana) in any significant way different from those that formed at the continental margins (as in the Ordovician Period in Vermont)? These questions suggest that however intrinsically interesting lateral facies studies at a given point in geologic time may be, they must be integrated *through* time if basic paleoecologic relationships and processes independent of time are to be clarified, and if the evolution of similar communities of organisms is to be defined.

Inherent Bias of the Fossil Record

As we discussed earlier in Chapter 3, organisms that secrete hard parts have a much greater opportunity for fossilization than soft-bodied organisms. Hence, a fossil assemblage usually contains only the remains of the former. Consequently, a fossil assemblage is but a small sample of the total original assemblage of species. Norman D. Newell, an American paleontologist, has estimated that, of some 10,000 species of organisms inhabiting a river-bank environment, only 10 to 15 are likely to be preserved as fossils in the river alluvium. For a coral reef environment he estimates that only 50 to 75 species will be fossilized out of a possible 3,000. Yet, as Newell points out, even the few species that are preserved in the fossil state, together with their sedimentary rock matrix, provide significant information about the original environment.

Besides the preferential preservation of hard parts, there is another important source of bias in the fossilization process—the differential preservation of environments depending on whether an area is being eroded or is being buried with sediments. Table 5–2 summarizes the importance of hard parts and rapid burial in providing a good record of ancient life. For example, an assemblage of organisms living in an upland prairie is less likely to be buried and preserved in the rock record than an assemblage that lives on the continental shelf, because the upland prairie environment is being more actively eroded than is the continental shelf environment. As the land areas are being eroded away, the submarine areas are subsiding and their sediments are accumulating. Because of this selective preservation of habitats, it is not surprising, therefore, that the fossil record of

Table 5–2

Factors Determining
the Fossil Record

| | | Hard Parts | |
		Yes	No
Rapid Burial	Yes	GOOD clams	POOR jellyfish
	No	POOR birds	NONE butterflies

marine shelly invertebrate assemblages is many times more complete than that of assemblages inhabiting land areas of high elevation such as prairies and plateaus.

Environmental Reconstruction

An interesting test case of the reliability of the fossil record, given its inherent bias for differential fossilization, was studied by Richard Konizeski, a vertebrate paleontologist, using fossil and Recent mammals from western Montana (Tables 5–3 and 5–4 and Fig. 5–3). What he did was this. First, he took a census of the

Table 5–3

Niches and Population for Recent Mammals*

Animal	Food Habits	Population Density (d)	Terrace p	Terrace dp	Flood Plain p	Flood Plain dp	Riparian p	Riparian dp	Coniferous p	Coniferous dp
Pronghorn	Large grazer	2	9	18	7	14				
Bison	Large grazer	2	6	12	8	16			2	4
Rocky Mountain sheep	Large grazer	1	8	8	8	8				
Whitetail jack rabbit	Small grazer	3	8	24	6	18			2	6
Cottontail rabbit	Small grazer	3	5	15	6	18			5	15
Mule deer	Large browser	2	3	6	5	10			8	16
Whitetail deer	Large browser	1	3	3	8	8			5	5
Elk	Large browser	1	6	6	7	7			3	3
Moose	Large aquatic plant-eater	1			7	7	6	6	3	3
Muskrat	Small aquatic omnivore	2					16	32		
Beaver	Small aquatic bark-eater	1			2	2	12	12	2	2
Pocket gopher	Very small fossorial granivore	3	5	15	8	24			3	9
Ground squirrel	Very small semifossorial granivore	3	6	18	6	18			4	12
Ground squirrel	Very small semifossorial granivore	3	6	18	6	18			4	12
Total = 14 species		s	11		13		3		11	
		t (dp)	133		174		50		91	
		% t (dp)	29%		39%		11%		20%	

d—total species population density weighted from 1 (low) to 3 (high)
p—species habitat-zone preference weighted from 0 (none) to 16 (total)
dp—species population density within a particular habitat zone
s—species representation, i.e., number of species occurring in a particular habitat zone
t—habitat-zone population density relative to total population density
% t (dp)—percentage of habitat-zone population density relative to total population density

From R. L. Konizeski, 1957, *Geol. Soc. America Bull.*, v. 68, pp. 131–150.

*The table shows the ecologic niches and population densities in river valley habitats for Recent mammals likely to be fossilized. Species in italics are those forms in the Pliocene Epoch which have been replaced ecologically by different modern species. Note the close similarity between predicted values of %t(dp) for Recent mammals and observed values from Pliocene fossils (Table 5–4).

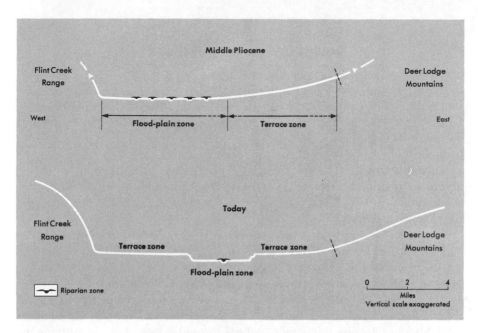

Figure 5–3 *Inferred distribution of Middle Pliocene habitats (top) and generalized distribution of the same habitats today (bottom). (From Konizeski, 1957.)*

Table 5–4

Niches and Population for Pliocene Mammals*

Animal	Food Habits	Population Density (d)	Habitat Zones							
			Terrace		Flood Plain		Riparian		Coniferous	
			p	dp	p	dp	p	dp	p	dp
Pronghorn ⎫		2	9	18	7	14				
Horse ⎬	Large grazer	2	9	18	7	14				
Horse ⎮		2	9	18	7	14				
Camel ⎭		2	9	18	7	14				
Camel	Large browser-grazer	2	7	14	9	18				
Rabbit	Small grazer	3	8	24	8	24			2	6
Peccary ⎫	Large rooter	2			8	16	4	8	4	8
Peccary ⎭		2			8	16	2	4	6	12
Proboscidean	Very large aquatic plant-eater	1			6	6	10	10		
Rhinoceros	Large aquatic plant-eater	1			6	6	10	10		
Beaver	Small bark-eater	1			1	1	14	14	1	1
Ground squirrel ⎫	Very small semifossorial granivore	3	8	24	2	6			6	18
Ground squirrel ⎭		3	8	24	2	6			6	18
Total = 13 species, 111 specimens		s	8		13		5		6	
		t (dp)	158		156		46		63	
		% t (dp)	37%		37%		11%		15%	

*The table shows the ecologic niches and population densities in river valley habitats for Middle Pliocene fossil mammals. All footnotes to Table 5–3 apply to this table.

Environmental analysis

present-day mammalian assemblage living in and around a small river valley that lies between two mountain ranges. Making allowance for changes in the fauna due to the effects of twentieth-century civilization (hunting, trapping, farming, spraying of pesticides, and so on), he found that there are in the area 49 species of mammals, including beaver, squirrels, bears, wolves, mice, rabbits, badgers, and elk. Of these 49 species, he conjectured, only 14 were likely to be preserved as fossils. The remaining 35 species, which he thought would be absent or poorly represented in a future fossil assemblage, either were very small, nonburrowing forms (various mice, for example) or had few representatives (bears and mink).

Konizeski next estimated the population density—high, medium, or low—and habitat preference—river terrace, flood plain, river bank or conifer forest—for each of the 14 species. In this way he was able to predict the probability of finding any given species as a fossil in a specific habitat.

Having established the predicted frequency of these mammals in various river valley and associated sediments he then collected the mammalian fossils from Middle Pliocene valley sediments of the same area. After identifying the fossil mammals and estimating their density in the different Pliocene habitats as inferred from the sedimentary rock matrix, he found a striking similarity between the Pliocene mammal community and the Recent one. Not only were the Pliocene and Recent habitats analogous in kinds and numbers but the relative abundance of species in terms of adaptive types was essentially similar. Konizeski also discovered that the ecologic niches filled during the Pliocene

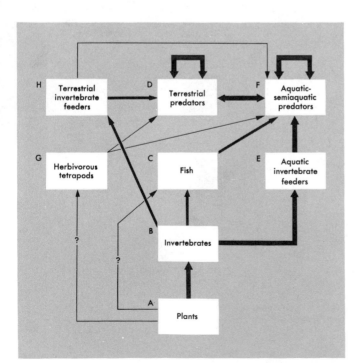

FIGURE 5–4 *The food chain in Early Permian times. The direction and strength of arrows indicate the relative food flow from one energy level to the next. A—Aquatic and terrestrial plants. B—Clams, snails, worms, insects, and other arthropods. C—Paleoniscoid fish. D—Carnivorous reptiles. E—Amphibians. F—Large amphibians, sharks, and lobe-finned fish. G—Herbivorous reptiles. H—Small carnivorous and omnivorous reptiles. (After E. C. Olson, 1961.)*

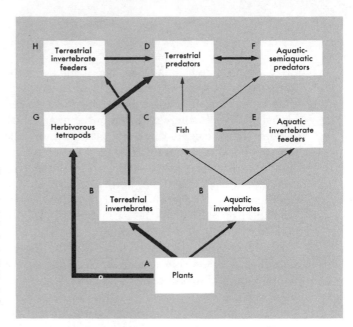

FIGURE 5-5 *Food chain in the Early Triassic Period. Symbols are the same as in Fig. 5-4. Note the shift in food flow from the aquatic and terrestrial invertebrate feeders to herbivorous reptiles. Primitive insectivorous mammals probably evolved from the terrestrial, invertebrate-feeding reptiles. (After E. C. Olson, 1961.)*

by horses, camels, rhinoceroses, and proboscideans are today filled by ecologically similar forms such as deer, elk, and moose. The most significant conclusion of this study is that a fossil community can be reconstructed from some, not necessarily all, representatives of the fauna and flora. This conclusion reaffirms what many paleontologists have intuitively felt to be true all along, namely that despite the loss of many members of an original assemblage of organisms, the few remaining fossilized species are usually adequate for environmental reconstruction.

This example has demonstrated that the ecologic roles that Pliocene and Recent mammals played in river valley environments has remained essentially the same. The "actors" filling some of the roles, however, have in some instances changed, although the roles remained the same. Thus, the ecologic role of river-bank-dwelling aquatic-plant-feeder was filled by proboscideans and rhinoceroses in the Pliocene; today that same role is filled by a different taxonomic group, a species of moose.

We might reasonably ask, however, if the structure of a particular community of organisms in a given habitat has ever changed with time. That is, have the ecologic roles as well as the organisms filling those roles ever varied significantly during Earth history? The answer to this question seems certainly to be affirmative, even though the problem of community evolution has only been recently approached by paleontologists.

An example of changes in community structure of Late Paleozoic and Early Mesozoic terrestrial reptiles has been studied by E. C. Olson. (See Figs. 5-4 and 5-5.) Olson noticed, as did other vertebrate paleontologists, that among reptilian genera in Late Paleozoic rocks, the ratios of carnivore to noncarnivore varied from 4:1 to 7:1. This observation is virtually the opposite of what we see today for terrestrial mammals, where the predators are outnumbered by the nonpredatory herbivorous forms. Olson restudied the Late Paleozoic vertebrate faunas and found that, although there were, indeed, few genera of terrestrial herbivores, there were a number of aquatic and terrestrial invertebrate-feeders. He concluded, therefore, that the Late Paleozoic terrestrial carnivores had varied food sources that included not only terrestrial herbivores but also a significant proportion of these invertebrate-feeders (Fig. 5–4). Gradually, however, a more varied group of herbivores evolved so that by the Early Mesozoic Era there were a larger number of terrestrial, vertebrate herbivores, forming a major part of the diet of the carnivores (Fig. 5–5). Olson points out that this alteration in the food-chain structure changes the composition of vertebrate communities from the Late Paleozoic to the Early Mesozoic. He believes, further, that the decreased dependence of these reptiles on water life (aquatic plants→aquatic invertebrates→aquatic and terrestrial invertebrate-feeders) made possible the adaptive radiation of these reptiles into a variety of upland, terrestrial environments. Olson also thinks that the origin of mammals from some of these reptilian groups was initiated by this major change in community structure.

6

Ancient environments and historical geology

Having discussed the interactions of organisms, sediments, and environments as well as the methods and assumptions of environmental analysis, we will now turn to some examples of ancient environments, showing how they are integrated into the broader field of historical geology. These examples are selected in order to demonstrate some of the wide variation that exists among ancient environments both in their paleo-ecologic setting and in the composition of their associated fossils. These examples will also indicate how variable the analysis itself might be, in terms of research techniques, kinds of information sought, and level of confidence in the conclusions drawn.

The Pleistocene-Recent Epochs of the North Atlantic

The North Atlantic Ocean basin is bounded by the northern portion of South America, the Caribbean islands, eastern

North America, the islands of Greenland and Iceland, the British Isles, western Europe, and the northwestern part of Africa. The basin connects directly with the South Atlantic through a relatively constricted area between the eastern and western continental projections of South America and Africa, respectively. Roughly circular in shape, the North Atlantic basin measures some 6,400 kilometers in diameter.

Although in several broad regions the North Atlantic reaches depths greater than 5,000 meters—almost twice that in the Puerto Rican trench—the bottom relief of the basin is quite variable. In fact, the variability of the topographic relief of the North Atlantic basin (and other ocean basins as well) is at least as great as that seen on the continental land masses. But despite this variability, however, certain physiographic provinces are evident (Figs. 6–1 and 6–2).

Much of the coarser, land-derived detritus that is brought to the oceans accumulates on the continental shelves, which are rather wide areas of little relief that extend seaward from land down to about 200 meters. The finer fraction is gradually winnowed out and carried to the deeper water areas beyond the shelves. Some of the fine-grained fraction that is deposited beyond the continental shelf may also include wind-blown material that has been carried far out to sea where it settles out onto the ocean surface. In addition, turbidity currents will occasionally transport relatively coarse sediment from the shelf edge, or down the submarine canyons that incise the shelf, and deposit it out on the abyssal plains. The sediments at the ocean floor, although variable, average some 500 to 1,000 meters in thickness and are as old as the Cretaceous Period.

Besides these inorganically derived sediments, there are also significant amounts of organic debris accumulating in the deeper parts of the ocean away from the continental shelves. This organic debris, which ordinarily would be greatly diluted by higher rates of inorganic sedimentation than are found in the open ocean, comes from a myriad of minute, calcareous tests secreted by floating protozoans, planktonic foraminiferans, as well as planktonic algae, or coccolithophorids. Other, less important skeletal sediments include the tests of planktonic snails (pteropods), various calcareous hard-parts of shelly invertebrates, fish bones, and so on.

FIGURE 6–1   *North Atlantic ocean basin showing major physiographic divisions. Symbols indicate location and dominant composition of cores collected from sediments of the ocean floor. The line across the middle of the figure shows the approximate position of the cross-section illustrated in Fig. 6–2. The insert indicates the approximate distribution of terrigenous muds (light gray) and calcareous muds (dark gray). (From D. Ericson and G. Wollin, 1964, and G. Arrhenius, 1963.)*

FIGURE 6–2   *(Far right) Topographic profile across the North Atlantic ocean basin showing major physiographic subdivisions. Note variation in bottom relief. Vertical scale is exaggerated so that slopes appear much greater than they actually are. 1,000 fathoms equal 1,829 meters. (From D. Ericson and G. Wollin, 1964.)*

Ancient environments and historical geology

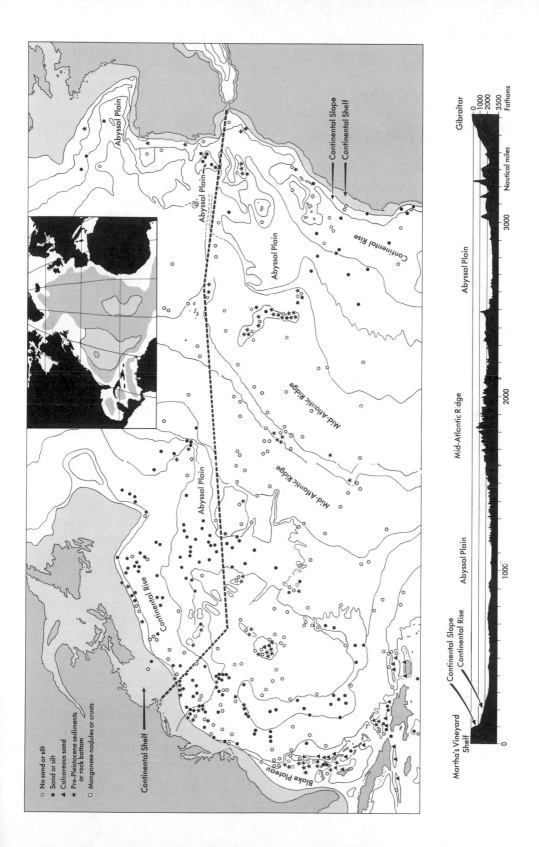

No sand or silt
Sand or silt
Calcareous sand
Pre-Pleistocene sediments or rock bottom
Manganese nodules or crusts

Continental Shelf

Continental Rise

Abyssal Plain

Mid-Atlantic Ridge

Mid-Atlantic Ridge

Abyssal Plain

Abyssal Plain

Continental Rise

Continental Slope
Continental Shelf

Blake Plateau

Martha's Vineyard
Shelf
Continental Slope
Continental Rise

Abyssal Plain

Mid-Atlantic Ridge

Abyssal Plain

Continental Rise
Continental Slope
Continental Shelf
Gibraltar

0
1000
2000
3500
Fathoms

1000
2000
3000
Nautical miles

Relative Chronology

With this brief introduction to the present-day North Atlantic basin, let us now consider the geologic history of the area during the Pleistocene Epoch, an interval of time spanning the last 2½ million years of Earth history.

From a study of surface features on the continents, such as erosional effects of ice and layers of glacially deposited sediments, geologists have been able to determine that, during most of the Pleistocene Epoch, a series of glaciations has occurred. Until fairly recently there was no substantial information about how these glacial events may have been recorded in the oceans—if at all. It was inferred, however, that sea level was lowered many tens of meters during glaciation because large quantities of water were bound up in the thick masses of ice that covered large portions of the northern hemisphere. During warm intervals sea level rose as this water was released to the world's oceans.

Extensive exploration of the oceans began after the turn of this century. Many data were collected on the character and circulation patterns of sea water, the topography and composition of the sea floor, the nature and thickness of deep-sea sediments, and the abundance and distribution of various marine organisms. Among the various oceanographic institutions engaged in this research, the Lamont Geological Observatory of Columbia University paid particular attention to the North Atlantic basin. The discussion that follows is based largely on the published results of Lamont's study of the deep-sea sediments of this area.

Part of the Lamont research involved collecting sediment cores from a large number of stations throughout the North Atlantic. These cores of soft sediment, most of which are 10 meters or more in length, provide a stratigraphic record of deep-sea geologic history. Although most of the cores record only the Recent and Pleistocene interval, about 10 per cent sample pre-Pleistocene sediments, some even going as far back as the Cretaceous Period. These older sediments at or near the seafloor's surface are due to the nondeposition or erosion of younger sediments.

Examination of the cores shows that they are mainly composed of interbedded layers of calcareous silt and sand, argillaceous mud, and quartz sand and silt. In some cores the layering or stratification is intact. In others, the stratification is badly disrupted by organisms that have burrowed through the sediments.

The calcareous layers are composed of the minute calcium carbonate skeletons of open-ocean planktonic protistan Foraminifera as well as of some calcareous Coccolithophorids (which are single-celled algae). In addition, some of the calcareous layers contain coarser, skeletal debris of shallow-water algae and shelly invertebrates, indicating that at times skeletal sediments have been transported from shallow-water environments to the deep ocean floor, presumably by turbidity currents. The quartz sand and silt layers represent transported grains from shallower, shelf areas into deep water, also by turbidity currents.

The argillaceous layers are composed of very small clay particles that have been transported from the lands far out to sea either in suspension in the upper waters or as wind-blown dust. These clay particles then slowly settle out of suspension and accumulate on the ocean bottom.

Argillaceous material is being deposited throughout the North Atlantic. The planktonic Foraminifera, too, are everywhere contributing to the deep-sea sediments. The relative rate of accumulation, however, will determine whether a given layer of sediment accumulating on the sea floor will be dominantly composed of clay particles or calcareous foraminiferal tests. Clays will predominate in areas close to major river systems or deserts. Elsewhere, the rate of production of organic calcium carbonate will swamp the contribution of clay. In waters deeper than 5,000 meters, however, although many foraminiferal shells may be drifting downward from the surface, they dissolve as they fall through the water and at the sediment surface, thus reducing their net contribution. Consequently, although calcareous foraminiferal sediments are being deposited throughout the North Atlantic beyond the continental shelves, they are not being preserved in the deeper portions of the North American basin, off the southeastern United States, nor in the deeper parts of the Canary and Cape Verde basins off the western coast of Africa (Fig. 6–1).

Because of the abundance of planktonic Foraminifera in these cores and because these single-celled organisms are sensitive to variations in water temperature, it was quickly realized that differences in the foraminiferal species in the cores might provide clues about the ancient temperature of the North Atlantic, particularly during the Pleistocene Epoch, when significant warming and cooling of the Earth's surface occurred.

The scientists studying these cores reasoned that first it would be necessary to determine which planktonic foraminiferans were accumulating in the ocean sediments today, a "warm" interval. Obviously, benthonic foraminiferans would not be helpful, for their distribution, if it is related to temperature, would be controlled by the temperatures prevailing at several thousand meters rather than by surface temperature. And yet surface temperature, of course, would be far more sensitive to major climatic changes than would the bottom waters of the North Atlantic.

By defining the planktonic foraminiferal composition of the uppermost layer of the cores, it would be possible to establish a reference point, or environmental datum, with which other, older assemblages might be compared. For example, one foraminiferan species, *Globorotalia menardii*, is a useful indicator for determining surface temperature because this species, which fluctuates in abundance in the calcareous layers of the cores, is strongly influenced by temperature. In the uppermost layers of the North Atlantic cores *G. menardii* is abundant; going down the core, however, this species disappears for a time, and then reappears once again. This variation in *G. menardii* had already been recognized by other scientists before the Lamont group. It was attributed to

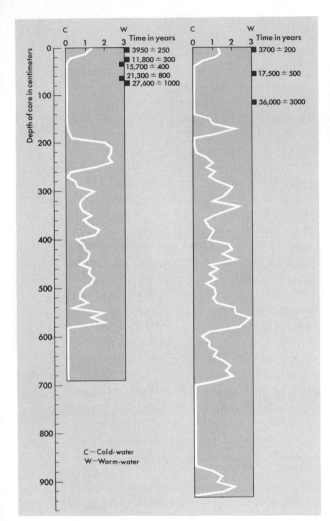

FIGURE 6–3 *Variations in abundance of the foraminiferan,* Globorotalia menardii, *in two cores from the Caribbean. The curves record changing ratios of G.* menardii *to weight of sediment coarser than 74 microns; low ratios indicate colder water temperatures, higher ratios, warmer water (ratios shown at top of columns). Cold intervals do not occur at same exact depth in the cores owing to relatively higher rates of sedimentation in the core at right than in that at left. Dates to right of each core were obtained by C14 method; note that since 10 to 15 thousand years ago the Caribbean has become warmer. Photograph shows several specimens of* Globorotalia menardii *enlarged about 11 diameters. (Drawing from Ericson and others, 1961; photo from Ericson and Wollin, 1964.)*

the end of the last glacial age and the beginning of the Recent episode of marine sedimentation and climatic amelioration. To judge from the *G. menardii* foraminiferal populations, therefore, a warm, preglacial interval was followed by a cold period (glacial interval) without *G. menardii*, which was followed by another warm period (postglacial) with *G. menardii* again (Fig. 6–3).

Besides *G. menardii*, some 15 to 20 other planktonic foraminiferal species and subspecies are used to define the deep-sea Pleistocene and Recent stratigraphy of the North Atlantic. But, once having established the sequence of foraminiferal assemblages in any one core, it is necessary to correlate it with the sequences established in other cores. The reason for this is that none of the cores contains a continuous record of the Pleistocene Epoch. These hiatuses are due to submarine slumping or submarine erosion by turbidity currents.

One useful method for correlating cores is based on a surprising phenome-

FIGURE 6–4 *Correlation between coiling direction of* Globigerina pachyderma *and surface water temperature. This foraminiferan coils to the left in colder waters and to the right in warmer waters. The boundary between the two different populations parallels the 7.2°C April isotherm. (From Ericson and Wollin, 1964.)*

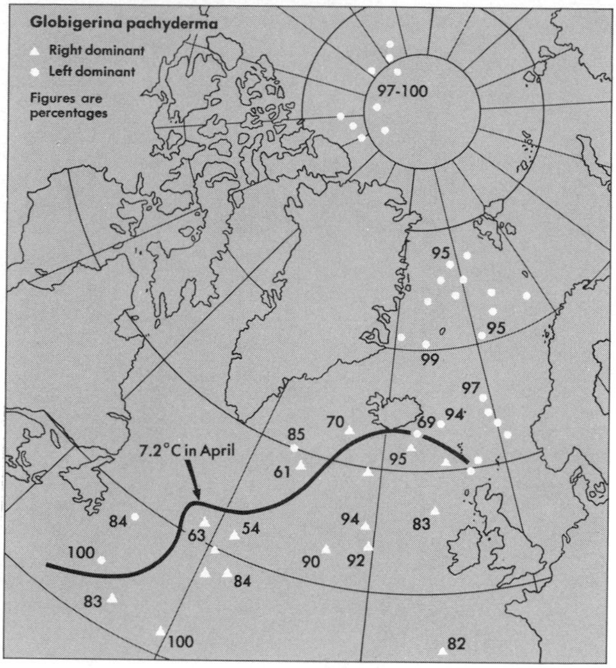

non: Some foraminiferal species apparently change their direction of coiling with changes in water temperature. For example, *Globigerina pachyderma*, the only planktonic foraminiferan living in the Arctic Ocean, coils to the left. But farther south, in subarctic waters of the North Atlantic it coils to the right. In this species, at least, coiling direction is dependent on water temperature (Fig. 6–4). Unfortunately, individuals of *G. pachyderma* are too few in most North Atlantic cores to use the reversal of coiling direction in this species to determine changes in Pleistocene temperatures. There is, however, a second abundant species, *Globorotalia truncatulinoides*, which exhibits similar changes in coiling direction. Although the temperature-dependent relationship in this species has not been definitely established, it is indicated because a similar phenomenon is observed in *Globigerina pachyderma* and because there is a change in coiling direction in *Globorotalia truncatulinoides* at the same stratigraphic level where *Globorotalia menardii* (the "warm"-water species) disappears. Thus, by noting the positions of coiling reversals in individual cores, it is possible to correlate a given stratigraphic level from one core to another throughout the North Atlantic, thereby permitting a reconstruction of the stratigraphic record from the Recent back through the Pleistocene and even into the Pliocene Epoch (Figure 6–5).

Further support for the conclusion that relative abundance of *Globorotalia menardii* and coiling direction of *Globorotalia truncatulinoides* mark changes in Pleistocene water temperatures of the North Atlantic is provided by oxygen isotopes. It has been observed that the relative amount in sea water of the two isotopes of oxygen, O^{18} and O^{16}, varies with water temperature. Planktonic

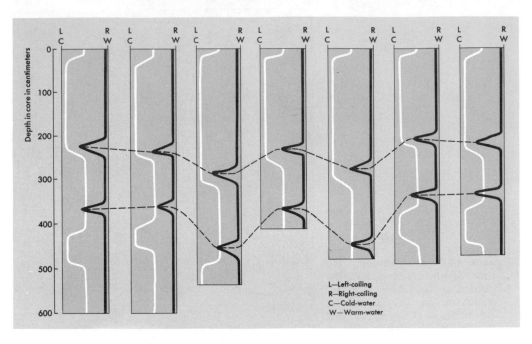

FIGURE 6–5 *Seven cores from the southern portion of the North Atlantic ocean showing generalized curves for cold-water and warm-water foraminiferan assemblages (white curve) and coiling-direction changes in* Globorotalia truncatulinoides *(black curve). Correlation of different stratigraphic levels in these cores is based on coiling-direction reversals; two separate time-correlative horizons are indicated by the two dashed lines connecting this series of cores. (From Ericson and Wollin, 1964).*

foraminiferans in secreting their $CaCO_3$ shells use the oxygen isotopes in the same proportion as in the surrounding sea water. Thus, shells secreted in colder water have a relatively higher ratio of O^{18} to O^{16} than do tests secreted by these same organisms in warmer water. The occurrence of these isotopic relationships in foraminiferans in North Atlantic cores agrees with the warm-cold intervals determined by *Globorotalia menardii* abundance ratios and by reversals in coiling direction of *Globorotalia truncatulinoides* (Fig. 6–6).

If we go back farther than the last warm preglacial interval, however, we find increasingly greater discrepancies between the data on oxygen isotopes and the data on abundance and coiling direction; scientists disagree about which set of data to believe. Doubt about the full validity for *absolute* temperature determination of the oxygen isotope data has been recently raised by the discovery that the planktonic foraminiferans do not spend their total life cycle in the uppermost part of the water. Rather, they descend to some 500 to 1,000 meters and secrete thick layers of calcite over their shells. Afterwards, they reproduce, and migrate upward as juveniles into the photic zone (upper 100 meters). Water temperatures, of course, decrease with water depth so that migration through this wide depth range would result in a foraminiferan test with a considerably variable $O^{18}/^{16}$ ratio. Moreover, the average value for the test as a whole would be lower than if the organism spent its full life in the warmer, surface waters.

Absolute Chronology

Once the relative chronology of the warm and cold intervals of the Pleistocene and Recent Epochs is established, how can absolute dates be obtained for determining an absolute chronology in years? Initially, measurements of the amount of the carbon isotope C^{14} in foraminifera tests were made because, being an unstable isotope, it decays at a fixed and known rate to N^{14}. Hence, older and older foraminiferans contain less and less C^{14}. Since the rate of decay of C^{14} to N^{14} is relatively rapid, materials much older than about 35,000 years cannot be reliably dated. Using the C^{14} dating method, however, an absolute chronology can be established for the upper layers of the cores that lie within

FIGURE 6–6 *Comparison of climatic curves based on foraminiferan assemblages and oxygen isotope ratios from two cores. Absolute ages of cores also given. Curves show good agreement in upper portion but with increasing discrepancies in lower portion (especially in core at left). Cesare Emiliani of the University of Miami, who obtained the oxygen isotope ratios from* Globigerinoides sacculifera, *places greater reliability on the isotope data assigning different absolute ages to Pleistocene stratigraphy from that shown in Fig. 6–7.*

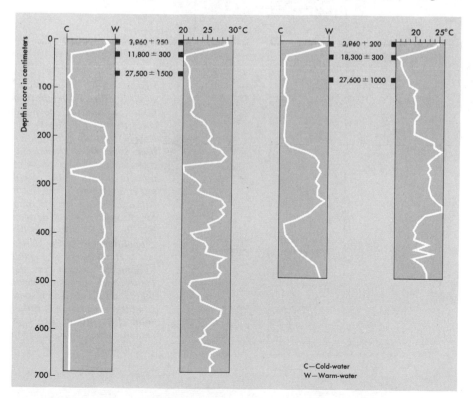

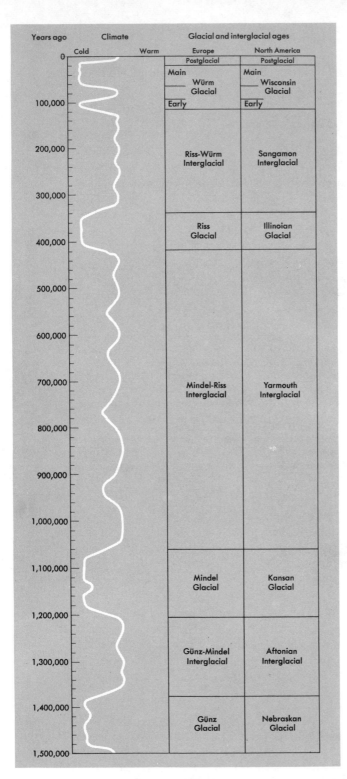

FIGURE 6–7 *Generalized climatic curve as determined from studies of deep-sea cores during the Pleistocene Epoch. Note that there have been four major glacial episodes separated by times of nonglaciation. Thus, deep-sea stratigraphy of the North Atlantic Ocean records similar climatic changes as seen on the continents in the glacial deposits of the Northern Hemisphere. Recent radiometric dates for the Pleistocene Epoch suggest that it may have begun some 2 million years ago rather than 1.5 million years as shown here. (From Ericson and Wollin, 1964.)*

this age range (Fig. 6–6). Interestingly, the last change from cold to warm in the cores, marking the end of the last glaciation, dates at about 11,000 years ago, which correlates with similar C^{14} dates obtained from continental Pleistocene-Recent stratigraphic sequences.

Other techniques are necessary to continue the absolute chronology for the lower parts of the cores that are too old for C^{14} dating. For instance, rates of sediment accumulation can be estimated by calculating the average thickness of sediments deposited in the last 30,000 years. Of course, due consideration must be made to eliminate from the calculation either cores with excessive rates of accumulation because of turbidity current deposition or cores with rates that are too low because of submarine slumping and erosion. Still, by judicious selection of appropriate cores where the sedimentation of particles falling through the water has been more or less continuous, a rate of accumulation of a few centimeters per thousand years seems likely. Using this figure, absolute ages can be approximated for the older parts of the Pleistocene record. More recent techniques have been developed for using the radioactive isotopes Protactinium^{231} and Thorium^{230} to obtain absolute ages directly. These radio-metric dates have confirmed the general validity of the ages determined by average rates of sedimentation and have provided a more refined absolute chronology for the last 300,000 years (Fig. 6–7).

Thus, we see how paleoecological interpretation of deep-sea sediments and their fossil faunas of Pleistocene age contributes to a broader interpretation of this time in Earth history. The waxing and waning of glaciation so vividly recorded in the northern latitudes of the continents   glacial scour and erosion, successive sedimentary deposits laid down by melt water, and coastline-terracing accompanying the rise and fall of sea level—are also recorded in deep-sea sediments, although more subtly.

The deep-sea environments are extremely constant, with temperature, salinity, and oxygen content remaining essentially the same for long intervals of time. Even in the oceans, however, climatic changes during glacial and interglacial periods brought about significant temperature variations. The temperature changes, in turn, exerted an influence on the distribution of the planktonic foraminiferans as well as altered their oxygen isotope ratios. These changes in temperature are thus recorded in the fossilized foraminiferan remains buried within the oceanic sediments.

One might suppose that much of the evidence for continental glaciation found on land today will be eventually lost through subsequent erosion. But the deep-sea evidence would not be so susceptible to loss and the record of the Pleistocene glaciation would therefore be kept intact, although it would certainly be less accessible.

The Cretaceous Period
of the Western Interior

During most of the Cretaceous Period, some 65 to 136 million years ago, the Western Interior of the North American continent was covered by a broad, shallow *epicontinental,* or *epeiric,* sea. From the distribution of marine and nonmarine Cretaceous rocks, we know that at the beginning of this chapter in geologic history the Western Interior was gradually encroached upon by seas transgressing northward from Mexico and southward from northwestern Canada. These two seas met in Wyoming toward the middle of the period and spread eastward, eventually forming a great seaway extending more than 6,000 kilometers in a north-south direction and 1,500 kilometers from the Rocky Mountains almost to the Mississippi River. Toward the end of the Cretaceous Period the sea retreated as gradually as it had appeared so that by the beginning of the Tertiary Period most of the continental interior was once again dry land. The continental margins along the Atlantic, Pacific, and Gulf Coasts were also inundated. Thus, at the maximum of the Cretaceous marine transgression, almost one-half of continental North America was covered by shallow seas (Fig. 6–8A).

Such transgressions and regressions of the seas as occurred in the Cretaceous Period were common throughout the geologic history of North America. The geosynclinal terrains around the periphery of the continent regularly subsided below sea level and received thick accumulations of sedimentary and volcanic material. At different times, each geosyncline underwent multiple mountain-building phases, or orogenies, resulting in the eventual formation of the Appalachian Mountains, Rocky Mountains, and Ouachita Mountains in the eastern, western, and south-central portions of the United States.

The structurally more stable continental interior also experienced these transgressions and regressions, and epicontinental seas covered parts or all of the continental interior, leaving a series of marine and nonmarine sedimentary rocks with many unconformities recording periods of nondeposition or erosion. Major marine transgressions culminated during the Late Cambrian-Early Ordovician, Late Ordovician-Early Silurian, Mississippian (Early Carboniferous), Pennsylvanian (Late Carboniferous), and Late Cretaceous Periods. Since the end of Cretaceous time, all the continental interior of North America and much of its border have been above sea level, undergoing nonmarine sedimentation or erosion.

The great Cretaceous sea was bordered on the west by highlands that had begun to form as a result of mountain-building activity during the Jurassic Period. These highlands, ancestors to the modern Rocky Mountains, experienced intermittent uplift during the Cretaceous Period and were the source of large volumes of sediment that entered, and ultimately filled, this sea. To the east,

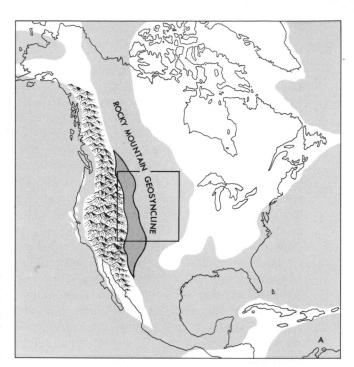

FIGURE 6–8 (A) Paleogeo-
graphic map showing distribution of lands and seas
during the time of maximum
marine transgression in the
Late Cretaceous Period. A
shallow sea covered the
Western Interior of North
America as well as Atlantic,
Pacific, and Gulf coastal
areas. Dark gray indicates
the area of nonmarine sediments being deposited from
erosion of the ancestral
Rocky Mountains. (From C.
O. Dunbar, 1960.) (B) Paleo-
lithologic map showing distribution of major types of
sediments within part of
Western Interior in Late Cretaceous Period. Dark gray =
nonmarine, lowland silts and
sands; stippling = nearshore
marine sands and silts; light
gray = offshore marine
muds. (From M. Kay and E.
Colbert, 1965.)

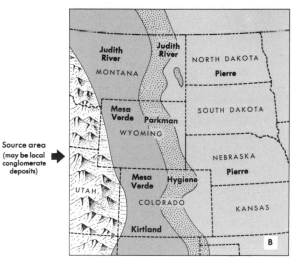

the lands were of very low relief and provided lesser amounts of relatively
fine-grained sediments (Fig. 6–8B).

The Cretaceous epicontinental sea advanced and retreated at rates that
would be virtually unnoticed by a human observer during the course of his lifetime. But over a span of several tens of millions of years, of course, the movement was of very great magnitude. The expansion of the sea correspondingly
altered the local environments with the Western Interior. Areas that were

Ancient environments and historical geology

initially being eroded began to receive nonmarine alluvial and fluvial sediments, then near-shore sands and silts, and eventually fine muds and, in places at the height of the transgression, calcium carbonate deposits. As the sea withdrew, these environments recurred in reverse order. Superposed on this idealized sequence of environments linked to the sea's regional migration were local fluctuations in depositional environment as great sheets of sediments intermittently flooded into the sea from the west as the ancestral Rocky Mountains underwent uplift and erosion. Consequently, the communities of terrestrial and marine organisms, in order to survive, were forced to migrate with the shifting and changing environments. Apparently, strong selection pressures were exerted on these organisms for, as we will see, many organisms became extinct during this time while others underwent rapid evolution.

There must also have been significant differences—in terms of salinity, oxygen content and nutrients—between the character of the epicontinental sea itself, and that of the open oceans, for many Cretaceous invertebrates that flourished in the other marine environments on the periphery of the North American continent and in other parts of the world were just about absent from the Western Interior sea (Table 6–1). As a broad generalization (because there are, indeed, many exceptions), we can say that the marine invertebrate assemblages of the Western Interior are thoroughly Mesozoic in character, lacking

Table 6–1

Cretaceous Fossils of the Western Interior*

Fossil Group	Record within Cretaceous Deposits of the Western Interior
FORAMINIFERANS	Benthonic and planktonic species common
Sponges	Unrecorded except for one shell-boring species and some spicules
Corals	Rare, a few small, solitary species; no reefs
Bryozoans	Locally abundant; very few species
Brachiopods	Nearly absent except for local inarticulates
CLAMS	Many species, inoceramids and oysters; most widespread and abundant
Snails	Less common than clams; locally very diverse
AMMONITES	Usually abundant, especially baculitids and scaphitids
Belemnoids	Abundant at a few localities; elsewhere absent
Echinoids	Very rare
Crinoids	Even fewer than echinoids
Crustaceans	Widespread, but abundant only locally

After J. B. Reeside Jr., 1957, *Paleoecology of the Cretaceous Seas of the Western Interior,* Geol. Soc. America, Memoir 67, v. 2, p. 512–513 and Donald Hattin, 1965, *Upper Cretaceous Stratigraphy, Paleontology, and Paleoecology of Western Kansas,* Geol. Soc. America Field Conference Guidebook.

*Relative abundances of marine invertebrate fossils in Cretaceous rocks of the Western Interior. Groups in capitals are particularly characteristic of the Western Interior; groups in italics have a poor record in the Western Interior but are common to abundant in Cretaceous marine deposits elsewhere in the world. These data indicate the relative lack of taxonomic diversity of the Cretaceous fossil assemblages of the Western Interior, suggesting an unusual ecology of the Western Interior seas as compared with the rest of the world.

many of the groups that, elsewhere, were beginning to appear in the Cretaceous Period and flourish in the Cenozoic Era. Although the Cretaceous marine deposits of the Atlantic and Gulf coasts contain abundantly diversified forms, those of the Western Interior, in general, show great endemism, or provincialism, and only a few groups are at all abundant.

Although it is, at present at least, difficult to define precisely what it was that was different between the seas of the Western Interior and those of the continental margins, certain conclusions seem reasonable. For example, we might expect that, because of the great distance from the open oceans, the Western Interior sea would have lacked good circulation with the normal marine waters of the Cretaceous oceans. Hence, any deviations in salinity (because of surface runoff from the surrounding land masses or because of excessive evaporation) or in oxygen content (because of reduction of organic matter) could not be corrected by periodic interchange with the large volume of ocean water because it lay so far away. In short, despite our present inability to specify what was different, it seems likely that the Western Interior sea was not, ecologically speaking, identical in character with the Cretaceous oceans. Consequently, marine invertebrate assemblages of the Western Interior represent communities adapted for life in this unusual epicontinental marine environment. More normal marine groups were generally restricted to the continental borders where environments were similar to those of the Cretaceous oceans. With the retreat of the Western Interior sea, the organisms living there became extinct, being unable to adapt and compete with the normal marine invertebrates.

An example of a stratigraphic sequence deposited by the Cretaceous epicontinental sea is found in western Kansas (Fig. 6–9). The summary of the late Cretaceous Period in this area, as given here, closely follows the paleoecologic study of this sequence made by Donald Hattin of the University of Indiana.

The major advance of the Cretaceous transgression reached western Kansas only about the beginning of the

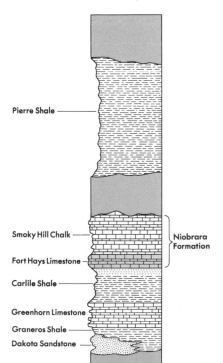

FIGURE 6–9 *Columnar section showing schematically the stratigraphy of the Upper Cretaceous in western Kansas. Total thickness is about 450 meters; individual formations are shown in proportionate thickness. Open gray shading indicates periods of nondeposition or erosion. (From R. C. Moore, 1958.)*

Pierre Shale

Smoky Hill Chalk — Niobrara Formation

Fort Hays Limestone

Carlile Shale

Greenhorn Limestone

Graneros Shale

Dakota Sandstone

second half of the period. Initial sediments were sands and silts with local coals representing deltaic deposits at the eastern margin of the transgressing sea. A complex of environments was associated with this delta, including channel deposits, flood-plain sediments, and lagoonal and marsh deposits, today represented by the Dakota Formation. Several varieties of clams and snails occur in the upper part of this stratigraphic unit. Interfingering laterally with the Dakota Formation is the Graneros Shale, which seems to record finer-grained sediments deposited in front (that is, seaward) of the deltaic complex. Fossils in the lower part of the Graneros—such as arenaceous benthonic foraminiferans, linguloid brachiopods, and certain clams, some of which are oyster-like—suggest waters of less than normal salinity. Discharge of surface water onto and across the delta apparently made the marine waters brackish here. There is in the Graneros an upward increase in the abundance and variety of fossils, such as ammonite cephalopods, clams and planktonic foraminiferans indicating gradually increased salinities toward more normal marine conditions.

As the transgression continued, the eastern margin of the sea moved farther eastward, where locally derived silts and sands continued to be deposited. In western Kansas, several hundred kilometers away, shallow water limestones began to accumulate as the lower Greenhorn Limestone. Later on, quieter waters prevailed with increased transgression so that the laminated chalks of the upper Greenhorn Limestone began to be deposited. At times the bottom conditions became unsuitable for invertebrates—perhaps owing to lowered oxygen content resulting from poor water circulation—for some parts of the Greenhorn Limestone lack benthonic fossils altogether (except for the ubiquitous inoceramid clams). Greenhorn fossil assemblages are dominated by ammonites, inoceramid clams, oysters, planktonic foraminiferans and some barnacles as well as numerous traces of burrowers.

Partial regression of the sea after the Greenhorn Limestone formed is reflected in the deposition of the next higher unit, the Carlile Shale, which locally becomes a quartzose silt and sand toward the top. Carlile fossils vary from an abundance of cephalopods, planktonic foraminiferans, and clams to a poorly fossiliferous assemblage near the top that includes scattered shark teeth, arenaceous benthonic foraminiferans, and rare molds of clams.

Renewed transgression is signaled by the deposition of the Niobrara Chalk, a fine-grained, relatively pure limestone derived from the comminution of shelly invertebrates, planktonic foraminiferans and coccoliths. As before, during the deposition of the Greenhorn Limestone, the bottom apparently temporarily stagnated and severely limited the establishment of a flourishing benthonic community. But at other times an abundant fauna existed, including the typical ammonite-inoceramid-oyster assemblage. A few additional elements such as belemnoids (squid-like cephalopods) and free-swimming crinoids, were also present (Fig. 6–10).

The beginning of the final regression of the Cretaceous epicontinental sea was marked by the deposition of the Pierre Shale, which represents the off-shore

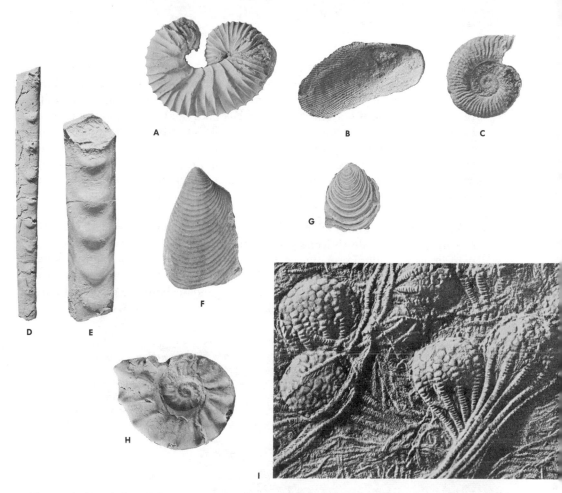

FIGURE 6–10 *A few of the more common marine invertebrates from the Upper Cretaceous of western Kansas. (A) Ammonite cephalopod,* Scaphites, *from the Carlile Shale, X ⅜. (B) Clam,* Brachidontes, *Upper Dakota Formation, X 1. (C) Ammonite cephalopod,* Watinoceras, *Greenhorn Limestone, X 1. (D) and (E) Ammonite cephalopod,* Baculites, *Pierre Shale, X ⅓. (F) and (G) two different species of the clam,* Inoceramus, *one from the Carlile Shale (F), X 1, and the other from the Greenhorn Limestone (G), X ¾. (H) Ammonite cephalopod,* Plesiacanthoceras, *from the Upper Graneros Shale, X ⅜. (I) Rock slab covered with well-preserved remains of the free-swimming crinoid,* Uintacrinus, *from the Niobrara Chalk, X ⅓. (A-C, F-H from D. Hattin, 1965; D-E from W. A. Cobban,* Journal of Paleontology *1962; I from R. C. Moore, 1958.)*

accumulation of muds coming from the erosion of lands far to the west and to the east. Fossils in the Pierre level, as in the strata below, are scarce in some parts of the unit and more abundant in others. Assemblages, when they do occur, are not very diverse, again being dominated by ammonites, inoceramid clams, and oysters. Total regression of the interior sea (and subsequent erosion) is presently marked in Kansas by the unconformity separating the Pierre Shale and the Upper Tertiary nonmarine deposits of Pliocence-Pleistocene age.

As noted earlier, and as the previous example indicates, Cretaceous marine invertebrate assemblages were not taxonomically diverse. These fossil assemblages are commonly dominated by one or few species of planktonic foraminiferans, swimming and bottom-feeding cephalopods, and benthonic inoceramid clams and oysters. Such low taxonomic diversity suggests an aberrant ecology of the Western Interior sea to which few organisms were able to adapt. The supposition of this marginal existence is supported by the fact that many stratigraphic horizons lack shelly invertebrates altogether, possibly indicating complete deterioration of the bottom environment to the extent that it could support no life at all.

Some evidence of mass killings of benthonic invertebrates has been provided by Karl Waagé of Yale University in his study of the fossiliferous concretionary layers in the Fox Hills Formation of the Late Cretaceous Period in South Dakota. Recurring concretionary layers with abundant individuals but few species of bottom dwelling invertebrates, mainly clams and scaphitid cephalopods, are

FIGURE 6–11 *Artist's reconstruction of a Late Cretaceous marine environment with a plesiosaur,* Elasmosaurus, *battling with a marine lizard,* Tylosaurus; *hovering overhead are the flying reptiles,* Pteranodon. *(Burian reconstruction.)*

separated by less fossiliferous strata in this formation. The concretionary layers are interpreted as recording periods of mass mortality of the benthonic fauna by excessive turbidity and lowered salinity resulting from the repeated influx of sediment-laden river water from land. The general absence in these concretionary layers of more mobile forms such as fish, belemnoids, and certain ammonites suggests that these swimmers were able to escape but that unfavorable changes in environment trapped and killed the more sluggish forms.

The marine Cretaceous deposits of the Western Interior also contain, in places, a rich record of the marine vertebrates of the time, especially fish and reptiles such as plesiosaurs, mosasaurs, and turtles (Fig. 6–11). These deposits also occasionally yield flying reptiles and early birds, which presumably fed on fish in the sea (Fig. 6–12).

Although life may have been somewhat limited and difficult in the Western Interior sea, it was extremely abundant and diverse on land. The flowering and fruit-bearing plants and trees, the *angiosperms*, which had appeared earlier in the Mesozoic Era, spread and flourished throughout the remainder of Cretaceous time. The land floras, which until then had been dominated by the less advanced *gymnosperms*, changed dramatically in composition and assumed a modern look.

Terrestrial reptiles, which were the dominant land vertebrates throughout the earlier Mesozoic Era, continued to flourish during the Cretaceous Period. A variety of herbivorous reptiles, including the heavily plated ankylosaurs, duck-billed ornithopods, and horned ceratopsians, were abundant. Carnivores such as the large predatory *Tyrannosaurus rex* presumably fed on these herbivores. Primitive mammals also were present as small, probably nocturnal carnivores, insectivores, and rodent-like forms.

As the North American continental interior began to emerge above sea level with the regression of the Late Cretaceous epicontinental sea, terrestrial habitats concomitantly increased in area and variety. Thus, an abundant and highly diverse terrestrial fauna and flora spread rapidly throughout the North American interior. From lowland swamps across wide alluvial plains and up into drier plateaus and foothills, Cretaceous animals and plants lived in a wide variety of environments forming many different terrestrial communities.

At the close of the Cretaceous Period, for reasons still unknown, many of these terrestrial organisms, which had flourished in many cases right up to the end, suddenly declined and became extinct. The gymnosperm plants were greatly reduced in numbers and habitats. Land dinosaurs died out. Flying and marine reptiles, too, became extinct. Other groups, however, such as angiosperms and mammals, increased even more in variety and numbers as they filled the ecologic niches left open by gymnosperm and reptilian extinctions (Fig. 6–13).

These extinctions at the end of the Cretaceous Period are even more puzzling than the peculiar nature of the Western Interior seas. Many theories have been advanced, but although they are perhaps adequate for explaining the ex-

FIGURE 6–12 (Left) Restored skeleton (courtesy Yale Peabody Museum), and (below) reconstruction of a Late Cretaceous diving bird from Kansas, Hesperornis. Note loss of wings and streamlined body; the total length of the animal is about 1.5 meters. (Courtesy American Museum of Natural History.)

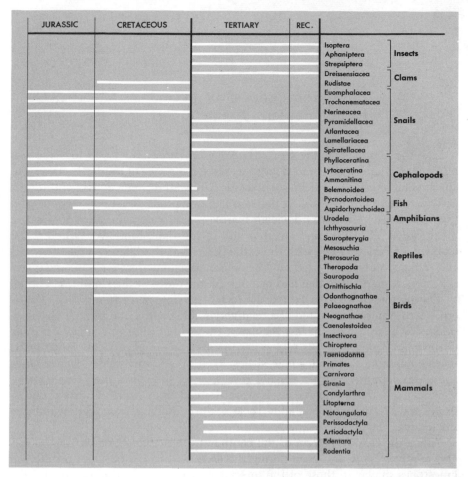

JURASSIC	CRETACEOUS	TERTIARY	REC.		
				Isoptera	Insects
				Aphaniptera	
				Strepsiptera	
				Dreissensiacea	Clams
				Rudistae	
				Euomphalacea	Snails
				Trochonematacea	
				Nerineacea	
				Pyramidellacea	
				Atlantacea	
				Lamellariacea	
				Spiratellacea	
				Phylloceratina	Cephalopods
				Lytoceratina	
				Ammonitina	
				Belemnoidea	
				Pycnodontoidea	Fish
				Aspidorhynchoidea	
				Urodela	Amphibians
				Ichthyosauria	Reptiles
				Sauropterygia	
				Mesosuchia	
				Pterosauria	
				Theropoda	
				Sauropoda	
				Ornithischia	
				Odonthognathae	Birds
				Palaeognathae	
				Neognathae	
				Caenolestoidea	Mammals
				Insectivora	
				Chiroptera	
				Taeniodontia	
				Primates	
				Carnivora	
				Sirenia	
				Condylarthra	
				Litopterna	
				Notoungulata	
				Perissodactyla	
				Artiodactyla	
				Edentata	
				Rodentia	

FIGURE 6–13 *The geologic ranges of a number of higher taxonomic categories of invertebrates and vertebrates. Extinctions at the end of the Cretaceous Period and new appearances in the beginning of the Tertiary resulted in a sharp biologic discontinuity between the Mesozoic and Cenozoic eras. Note the great ecologic and taxonomic variety represented among the extinct groups. (From N. D. Newell, 1962.)*

tinction of one group or another, they invariably fail to explain why so many different kinds of organisms met this fate. The Cretaceous extinctions cut across groups that are highly divergent ecologically and taxonomically. Was this great dying-out merely the fortuitous coincidence of many separate, unrelated extinctions? For after all, extinction of individual lineages had been occurring regularly throughout the Paleozoic and early Mesozoic Eras. Or was it due to some universal phenomenon that triggered the extinction of a number of key, ecologically critical groups on the land and in the sea, thereby upsetting the whole ecologic balance, terrestrial as well as marine, and causing other groups to also become extinct? The Cretaceous extinctions, like those at the end of the Paleozoic Era, pose one of the most significant challenges for present-day paleoecology.

The Ordovician Period
of Eastern North America

Like the Cretaceous, the Ordovician Period (from about 500 to some 430 million years ago) was a time of large-scale submergence of the North American continental interior by epicontinental seas. Unlike the Cretaceous counterparts, however, these Ordovician interior seas had numerous direct connections with the open oceans so that Ordovician marine deposits contain many fossils of a great variety of the marine invertebrates of the time. Moreover, whereas the Cretaceous Period closed with the extinctions of many different organisms, the end of the Ordovician Period saw the culmination of a great increase and diversification of marine life that had begun with the dawn of the Cambrian Period.

During the Ordovician Period the tectonic-sedimentary framework of eastern North America consisted of a relatively stable continental interior that subsided slowly, and more or less regularly, throughout most of Ordovician time. Ordovician deposits in this region are mainly composed of richly fossiliferous marine limestones and calcareous shales. Well-sorted, fine-grained quartz sands were also deposited during a major regression of the sea in the Middle Ordovician Period. These quartz sands were derived from earlier Cambrian sandstones and Precambrian crystalline rocks of the central interior and Precambrian shield areas.

Along the eastern margin of the continent, subsidence beneath the sea was also relatively slow and even during the early part of the period. Fossiliferous limestones similar to those of the interior region were deposited. Toward the end of the early part of the Ordovician Period, however, rates of subsidence increased considerably. Thick deposits of black shales and graywackes began to accumulate. These rocks represent the erosional debris entering the Ordovician sea from the east, as volcanic island arcs began to form within what today is known as the Appalachian Geosyncline (Fig. 6–14). The Appalachian Geosyncline, which was a broad, linear belt along the eastern border of North America, subsided greatly, climaxed by repeated uplift and deformation of its sedimentary and volcanic rocks throughout most of the Paleozoic Period. The first major orogenic phase developed in the second half of the Ordovician Period, culminating in the *Taconic Orogeny* late in the period. This mountain-building activity resulted in the accumulation and spread of several fluvial and deltaic complexes along the eastern third of the United States. Toward the continental interior these terrigenous deposits passed laterally into marine limestones and calcareous shales (Fig. 6–15).

The seas of the stable interior during the Ordovician Period were shallow, probably warm, and well-lit, and the land-derived sediments that entered were fine-grained and slowly deposited. Consequently, the sediments that accumu-

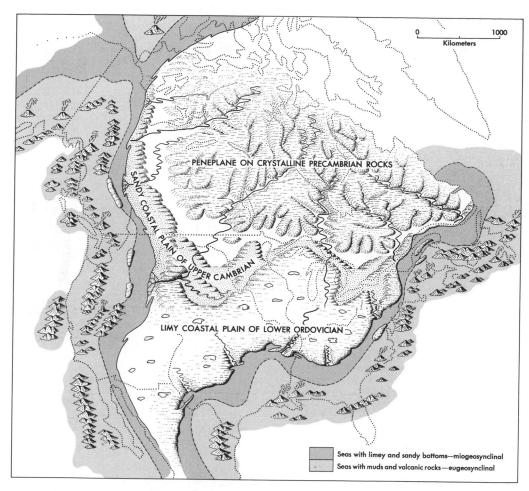

FIGURE 6–14 *Paleogeographic map of North America during the Middle (Chazyan) Ordovician Period. After transgressing the continent in Early Ordovician times, the seas retreated and then partially readvanced during the middle part of the period. During the Late Ordovician Period the continent was once more inundated. The continental margins are bordered by the tectonically active geosynclines which persisted through the Paleozoic Era in the east, and until the end of the Mesozoic Era in the west. (From M. Kay and E. Colbert, 1965.)*

lated were calcareous shales and shaly limestones rich in marine fossils. These rocks are termed *shelly facies* because of their abundant and diverse shelly invertebrates—corals, bryzoans, articulate brachiopods, and trilobites being especially common.

By contrast, the marine sediments of equivalent age laid down along the eastern continental margin within the geosyncline were usually (but not always)

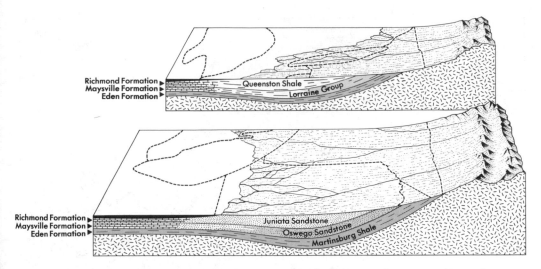

FIGURE 6–15 *Block diagram and cross section of eastern North America. (Pennsylvania and New York) at the end of the Ordovician Period. Deltaic and fluvial sediments have built out from the highland areas to the east which have been formed by the Late Ordovician Taconic Orogeny; farther west these nonmarine units pass into nearshore and offshore silts and limestones. Previous to the deposition of the nonmarine Queenston and Juniata sediments, marine shales and sandstones were deposited. As the tectonic uplift proceeded in the east, increasingly coarser silts and sands were laid down. During the Late Ordovician Period the sea's margin gradually migrated westward as the erosional debris from the east began to fill in the Appalachian sedimentary basin. Length of section about 1,000 kilometers. (From C. O. Dunbar, 1960.)*

deeper-water black shales and graywackes. These rocks are termed the *shaly facies* and have rather different fossil assemblages that are less abundant and less diverse than the shelly facies. Inarticulate brachiopods, certain trilobites, and graptolites are typical of these deposits (Fig. 6–16).

The differences in the fossils of the shelly and shaly facies are due to ecologic differences in the two major environments in which these rocks were deposited. The shallow, epicontinental seas of the interior supported a wide variety of organisms living in a number of different marine habitats. Environmental variety was undoubtedly provided by variations in bottom relief and water turbulence within this shallow sea. Minor fluctuations in sea level effected major environmental changes owing to the relatively shallow water depths involved.

The ecology of the geosynclinal belt, however, was markedly different. Here, water depths were usually great enough so that variations in bottom topography or sea level had less importance. Bottom conditions were probably relatively constant for long periods of time. Substrates were often oozy and stagnant, as during black shale deposition, so that only a few bottom-dwelling invertebrates could flourish.

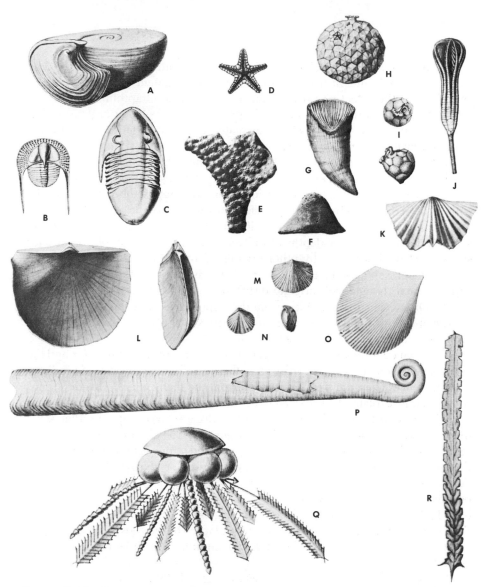

FIGURE 6–16   *Representatives of some common marine invertebrates from Ordovician rocks of North America. (A) Snail* Maclurites, *X* ⅗. *(B) Trilobite,* Cryptolithus, *X* ½. *(C) Trilobite,* Isotelus, *X* ⅙. *(D) Starfish,* Hudsonaster, *X* ⅗. *(E) Bryozoan,* Constellaria, *X* ⅗. *(F) Bryozoan,* Prasopora, *X* ⅗. *(G) Solitary rugose coral,* Streptelasma, *X* ⅗. *(H) Primitive attached echinoderm, the cystoid,* Echinosphaerites, *X* ⅗. *(I) Top and side views of the cystoid,* Malocystites, *X* ⅗. *(J) Crinoid,* Ectenocrinus, *X* ⅗. *(K) Brachiopod,* Platystrophia, *X* ⅗. *(L) Bottom and side views of the brachiopod,* Rafinesquina, *X* ⅗. *(M) Brachiopod,* Resserella, *X* ⅗. *(N) Side and bottom views of the brachiopod,* Zygospira, *X* ⅗. *(O) Clam,* Byssonychia, *X* ⅗. *(P) Nautiloid cephalopod,* Lituites, *X* ⅔. *(Q) Graptolite colony with float and reproductive "pouches,"* Diplograptus; *branches or stipes with individuals arrayed along stipes in small cup-like structures, slightly reduced. (R) Graptolite stipe of* Climacograptus, *lower part of stipe normal, upper part flattened, slightly reduced. (From C. O. Dunbar, 1960.)*

Not only were ecological factors important in initially differentiating the shelly facies environment from the shaly facies environment but they were also important in the selective preservation of graptolites. Graptolites, which are extinct today, are presumed to be early protochordate colonial organisms that had a delicate exoskeleton of chitinous material. Many varieties of graptolites in the Ordovician Period were planktonic in habit, so that when a colony died it slowly drifted down to the sea floor. In the shelly facies, graptolites are not usually preserved because either the exoskeleton was soon destroyed by scavengers and bottom currents, or else it oxidized and decomposed before final burial within the sediments. In the shaly facies, however, periodic stagnation of the bottom environment eliminated most if not all of the benthonic fauna. Here, in the quiet, oxygen-poor depths of the geosynclinal area, graptolites were far less subject to destruction by other organisms, oxidation, or current activity. Hence, the preservation of abundant graptolites is restricted to the geosynclinal deposits even though their original distribution was far more widespread during the Ordovician Period. Once again the importance of ecology in determining the abundance and distribution of organisms is illustrated, even though the ecologic influence, in this instance, is quite indirect.

When we consider the Ordovician fossils of the continental interior and the geosynclinal belt, it is noteworthy that there are on the average more long-ranging genera from the shaly facies than from the shelly facies. This difference is apparently explained by the fact that the geosynclinal environments were probably less variable and more constant for long intervals of time than were the shelly environments. Periodic fluctuations in sea level, often even complete regression of the interior seas, presumably exerted strong selection pressures on the marine invertebrates of the continental interior during the Ordovician Period. Consequently, evolutionary rates were high and many local species became extinct while new ones evolved. These sea level fluctuations would have had far less effect in the deeper-water geosynclinal belts, and any regressions there were probably quite local. Once adapted to the geosynclinal habitats, organisms would evolve little, instead maintaining the adaptations already achieved.

Because of the faunal differences between the shelly and shaly facies, it has often been difficult to correlate strata in one sequence with those in the other. Without similar fossils it is virtually impossible to determine precise stratigraphic relations between the two. For example, some 15 subdivisions of the Ordovician Period have been established in the British Isles by means of the chronologic sequence provided by the graptolite assemblages of the shaly facies there. Graptolite-bearing units in eastern North America, several thousand kilometers away, can usually be closely correlated with their British counterparts. But Ordovician limestones of the central interior of the United States cannot be so accurately correlated with North American shaly facies only hundreds or even tens of kilometers away.

Besides the lateral changes in Ordovician faunas seen as the shelly facies of the continental interior passing into the shaly facies of the continental margins, there are significant temporal changes in the faunas. Comparison of Early Ordovician fossiliferous rocks with Late Ordovician units shows that during the Ordovician Period the marine invertebrate faunas became increasingly more varied, and individual groups became come abundant. (As N. D. Newell has pointed out, the approximately 100 families of fossil organisms of the Late Cambrian Period expand to almost twice that number by the end of the Middle Ordovician Period.) This expansion and diversification of Ordovician marine invertebrates is a culmination of the *adaptive radiation* of all marine organisms that began in Early Cambrian and continued throughout Ordovician time (Table 6–2). By the end of the period all the phyla and some two-thirds of the classes of fossilizable marine animals that have appeared in Earth history were present. Post-Ordovician evolution, as complex and diverse as it is, involved "variations on the major themes" introduced in the Cambro-Ordovician seas. This post-Ordovician evolution consisted chiefly in exploiting new habitats (for instance, invasion of terrestrial environments) and improving adaptations in existing habitats (such as increased mobility and feeding mechanisms in fish). Origins of new groups and extinctions of existing groups occurred predominantly at the species and generic levels, and correspondingly less within increasingly higher taxonomic categories—families, orders, classes, and phyla (Fig. 6–17).

An interesting example of the expansion of shelly marine invertebrates in the Early Paleozoic seas is provided by the Middle Ordovician reef communities of western Vermont and eastern New York whose paleoecology has recently been studied by Max Pitcher. These reefs are revealing because they include many representatives of the newly evolved marine invertebrates making their

Table 6–2

Adaptive Radiation of Marine Organisms*

Period	Phyla	Classes
Today	12	31
Ordovician		
Late	12	33
Middle	12	32
Early	11	27
Cambrian		
Late	11	22
Middle	10	20
Early	8	12

From G. G. Simpson and W. Beck, 1965, *Life: An Introduction to Biology*, 2nd ed., N.Y., Harcourt, Brace & World, pp. 770.

* Numbers of marine phyla and classes as fossils in Cambrian and Ordovician rocks and those with fossilizable hardparts living today.

first appearance in the Ordovician Period, creatures such as stromatoporoid and tabulate corals and trepostome, cyclostome, and cryptostome bryozoans. Although earlier reef-like structures, formed by algae (stromatolites) or primitive sponge-like archaeocyathids are known in the Cambrian Period, it was not until Middle Ordovician time that corals became important reef-formers, as they have remained until today.

At the end of the Early (Canadian) Ordovician Period when shallow-water limestone had been deposited there was a regional regression of the seas in the New York-Vermont area. The Middle (Chazyan) Ordovician Period opened with a renewed transgression marked by the deposition of a fine-grained quartz sandstone, possibly a beach or barrier bar deposit, in the wake of the westward advancing sea. Seaward of this trangressive sandstone, fossiliferous marine limestone accumulated (Fig. 6–18).

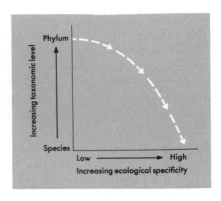

FIGURE 6–17 *Decreasing ecologic specificity with increasing taxonomic level. Changes in environment are thus more likely to "overstep" adaptions of lower taxonomic groups, which are specialized for narrowly defined ecologic niches, than of higher ones. Consequently, no phyla have become extinct since the end of the Ordovician Period, but increasingly more and more of the lower categories have, so that today no species, and possibly only a single genus,* Lingula, *an inarticulate brachiopod, survive from the Ordovician Period.*

In places small mounds up to several meters across and high grew up from the sea floor. Initially these mounds, or small reefs, were built by encrusting bryozoans which trapped calcareous mud and disarticulated fragments of primitive, attached echinoderms (cystoids) among the bryozoan colonies. Between these reef-like mounds calcareous sands accumulated, formed by the skeletal debris of ostracodes, trilobites, cystoids, and bryozoans that lived between the reefs. Scattered quartz silt and sand, together with abundant cross-stratification, suggest current agitation and proximity to land.

With time, these Early Chazyan reefy limestones continued to accumulate with variations in mound geometry (some showing regional alignment to existing current directions) and contributory organisms. In places large tabulate coral fragments are found as storm-tossed boulders encrusted by bryozoans.

Middle Chazyan reefs are composed of stromatoporoid corals and calcareous algae, which later give way to an assemblage dominated by tabulate corals, sponges, and other stromatoporoids (Fig. 6–19). This assemblage is contemporaneous with other, separate trepostome bryzoan reefs developing elsewhere in the area. A third common assemblage within the Middle Chazyan series is a shallower-water, perhaps intertidal, community of algal stromatolites, cal-

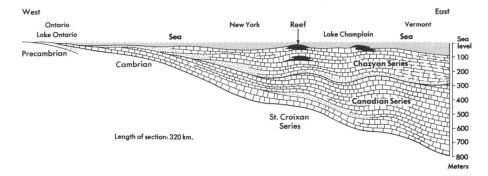

West   East
    Ontario   New York   Reef   Vermont
    Lake Ontario   Sea   Lake Champlain   Sea
Precambrian   Sea level
    Cambrian   Chazyan Series
                                                                                    Canadian Series
                          St. Croixan
                            Series
Length of section: 320 km.

FIGURE 6–18 *Restored section across northern New York and northwestern Vermont showing stratigraphy of Lower (Canadian) and Middle (Chazyan) Ordovician Period. Rocks thicken eastward away from stable interior into the western part of the Appalachian geosyncline. Quartz sandstone at the base of the Chazyan Series marks the transgression at the beginning of Chazyan following regression at the end of the Canadian period. Patch reefs developed offshore during Chazyan time. (From M. Kay and E. Colbert, 1965.)*

FIGURE 6–19 *Distribution of main reef-dwelling organisms in Early, Middle, and Late Chazyan time. The tabular and encrusting organisms were important in binding the reef sediments together into a rigid wave-resistant structure that could consequently grow up into wave-agitated surface waters rich in nutrients, oxygen, and sunlight. The nontabular and nonencrusting organisms were important in providing the frame-building materials bound together by the tabular and encrusting forms. Note the increase in diversity of the reef-dwelling organisms during Chazyan time, reflecting the adaptive radiation of marine shelly invertebrates that was taking place around the world during the Ordovician Period as a whole. (From M. Pitcher, 1964.)*

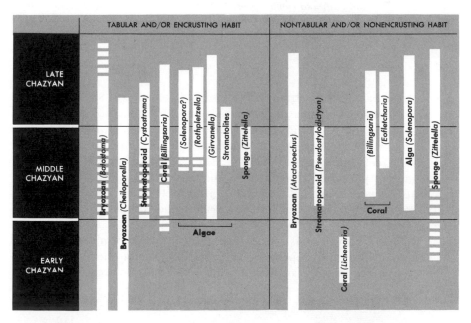

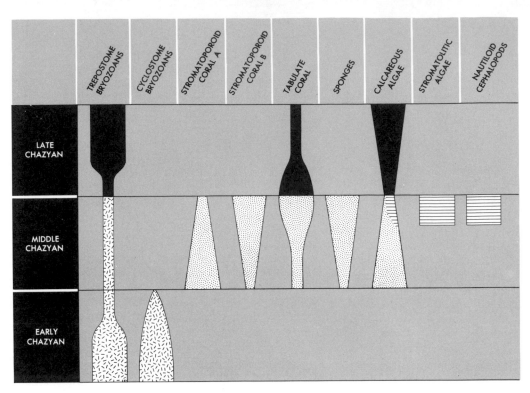

FIGURE 6-20 *Highly generalized distribution chart of the different dominant reef assemblages during Chazyan time. Trepostome and cyclostome bryozoan reefs were important in Early Chazyan time (dashed areas). The trepostomes continued as a separate reef assemblage in the Middle Chazyan Period; a second assemblage (dotted areas) appeared at that time, one which initially was dominated by stromatoporoids and calcareous algae but gradually was modified into an assemblage dominated by different stromatoporoids, tabulate corals, and sponges; a third Middle Chazyan assemblage (striped areas) was composed of calcareous algae, and nautiloid cephalopods. In Late Chazyan time, the trepostome byrozoans formed a single dominant reef assemblage (black areas) together with tabulate corals and calcareous algae. (Data summarized from M. Pitcher, 1964.)*

careous algae, and current-transported nautiloid cephalopod shells. The presence of various kinds of algae, oölites, algal-coated grains, tumbled coral heads, and erosional channels, within and near the reefs, indicate that the waters were shallow and well agitated, where abundant nutrients, oxygen, and sunlight allowed the reef communities to flourish. The encrusting habit of many of the reef organisms permitted the formation and accumulation of a rigid, mound-shaped framework that could tolerate considerable wave agitation, thereby enabling the reefs to grow up into the shallow and turbulent surface waters (Fig. 6–19).

The Upper Chazyan reefs are dominated by a single assemblage composed of calcareous algae and trepostome bryozoans. Here, for the first time, calcareous algae are directly associated with bryozoans. Before this time, calcareous algae had either been absent from the reef communities (as in Early Chazyan time) or else in a separate, contemporaneous, nonbryozoan, reef-building assemblage (Fig. 6–20).

As Pitcher states in his study of these reefs, a variety of shelly marine invertebrates invade and flourish in the reef habitat for the first time in the Middle Ordovician Period. Hereafter throughout the Paleozoic Era, many of these same organisms are important reef-formers and inhabitants. And even today, tropical reefs include the descendants of groups that first appeared in the Chazyan reefs, some 450 million years ago.

Summary

In this chapter we have seen how knowledge of ancient environments can be integrated with paleontology, stratigraphy, and sedimentation into a broad interpretation of Earth history. The examples that have been provided demonstrate the variety of interesting problems in historical geology that are inextricably intertwined with environments.

We have seen how the major climatic fluctuations of the Pleistocene Epoch are recorded in the fossil assemblages of deep-sea sediments, complementing and corroborating the evidence on the continents for periodic glaciation.

The Ordovician and Cretaceous Periods provide good examples of how the major environmental differences between seas lying within the continental interior and those along the continental margins result in markedly different marine fossil assemblages. These differences pose difficulties for the stratigrapher who relies on the similarities of fossil assemblages to correlate rocks in one region with those in another.

Changes in the distribution of environments, as discussed in each of the examples, result in expansion or contraction of the populations of organisms living in those environments. These variations in availability of living space undoubtedly exert differing selection pressures on the organisms involved, thereby influencing evolutionary rates expressed in origins of new groups and extinctions of existing ones.

We may conclude, therefore, with the realization that an adequate understanding of the stratigraphic record of sedimentary rocks, for explaining both the history of life and the temporal sequence and variation of sedimentary environments, requires knowledge of the nature of the ancient environments in which the sediments were deposited and the organisms flourished.

Suggestions for further reading

General

Beerbower, J., 1960, *Search for the Past,* Englewood Cliffs, N.J., Prentice-Hall, Inc.

Dunbar, C., 1960, *Historical Geology,* 2nd ed., New York, John Wiley and Sons.

Kay, M. and E. Colbert, 1965, *Stratigraphy and Life History,* New York, John Wiley and Sons.

Stokes, W. L., 1966, *Essentials of Earth History,* Englewood Cliffs, N.J., Prentice-Hall, Inc., 2nd ed.

Woodford, A. O., 1965, *Historical Geology,* San Francisco, W. H. Freeman.

Chapter 2
Sediments and environments

Dunbar, C. and J. Rodgers, 1957, *Principles of Stratigraphy,* New York, John Wiley and Sons.

Krumbein, W. and L. Sloss, 1963, *Stratigraphy and Sedimentation,* San Francisco, W. H. Freeman and Co., 2nd ed.

Pettijohn, F., 1957, *Sedimentary Rocks,* New York, Harper and Row, 2nd ed.

Weller, M., 1960, *Stratigraphic Principles and Practices,* New York, Harper and Row.

Chapter 3
Organisms and environments

Allee, W., A. Emerson, O. Park, T. Park, and K. Schmidt, 1949, *Principles of Animal Ecology,* Philadelphia, W. B. Saunders and Co.

Hedgpeth, J. (Editor), 1957, *Treatise on Marine Ecology and Paleoecology,* Geol. Soc., America Memoir 67, vol. 1.

MacGinitie, G. E., and N. MacGinitie, 1949, *Natural History of Marine Animals,* New York, McGraw-Hill.

Moore, H. B., 1958, *Marine Ecology,* New York, John Wiley and Sons.

Simpson, G. and W. Beck, 1965, *Life: an Introduction to Biology,* New York, Harcourt, Brace and World, 2nd ed.

Yonge, C. M., 1949, *The Sea Shore,* London, Collins.

Chapter 4
Geochemical environmental evidence

Degens, E. T., 1965, *Geochemistry of Sediments,* Englewood Cliffs, N.J., Prentice-Hall, Inc.

Mason, B., 1966, *Principles of Geochemistry,* New York, John Wiley and Sons, 3rd ed.

Chapter 5
Environmental analysis

Ager, D., 1963, *Principles of Paleoecology,* New York, McGraw-Hill.

Imbrie, J. and N. Newell (Editors), 1964, *Approaches to Paleoecology,* New York, John Wiley and Sons.

Chapter 6
Ancient environments and historical geology

Ericson, D. and G. Wollin, 1964, *The Deep and the Past,* New York, Alfred A. Knopf, Inc.

Ladd, H. (Editor), 1957, *Treatise on Marine Ecology and Paleoecology,* Geol. Soc. America Memoir, vol. 2.

Newell, N., J. Rigby, A. Fischer, A. Whiteman, J. Hikcox, J. Bradley, 1953, *The Permian Reef Complex of the Guadalupe Mountains Region, Texas, and New Mexico,* San Francisco, W. H. Freeman and Co.

Zangerl, R. and E. Richardson, 1963, *The Paleoecological History of Two Pennsylvanian Black Shales,* Fieldiana, Geology Memoirs, vol. 4, Chicago Natural History Museum.

Index

Fox Hills Formation, 98
Fusulinids, 34–35

Gametes, 33
Genotype, 31
Geochemistry:
  calcareous skeletons and, 62
  environments and, 61–68
  isotopes and, 63–66
Geosyncline, 25, 92, 102, 103, 104,
  106
  Appalachian, 102
Glaciation:
  Pleistocene, 84–91
  sediments and, 10
Globigerina pachyderma, 87
Globorotalia menardii, 85–88
Globorotalia truncatulinoides, 87–
  88
Graded bedding, 15
Grain cementation, 16
Graneros Shale, 96
Granite, weathering of, 8–9
Graptolites, 106
Great Bahama Bank, 24, 45, 72
Greenhorn Limestone, 96
Gulf Stream, 10–11
Gymnosperms, Cretaceous, 99

Habitats, 31
  selective preservation of, 75–76
Hadal region, of ocean, 4
Hall, James, 70
Hattin, Donald, 95
Herbivores, 48
Historical geology, 81, 111
  correlating strata (Ordovician),
    102–111
  Cretaceous extinctions, 92–101
  relative and absolute Pleistocene
    chronology, 84–91
Hutton, James, 70

Ice, sediments and, 10
Imbrie, John, 24
Industrial melanism, 43
Isotopes:
  carbon, 65–66, 89–91
  geochemistry and, 63–66
  N14, 89
  oxygen, 64–65, 87–88
  protactinium, 91
  thorium, 91

Konizeski, Richard, 76, 78

Lacustrine environment, 5
Lakes:
  oxygen and, 39
  sunlight and, 40
  water turbulence and, 41–43
Lamont Geological Observatory, 84
Lamprey, as parasite, 50
Larval stage, of marine inverte-
  brates, 33
Lateral facies studies, 73–75

Liebig, Justus, 53
Liebig's rule, 53
Life, antiquity of, 67
Limestone, 108
  Greenhorn, 96
Limiting factor, example of, 53
Lithofacies, 28
Littoral environment, 4
Lowenstam, Heinz, 62
Lyell, Charles, 70

Magnesium, water temperature
  and, 63
Mammals, teeth of, 34–35
Marine environments, 4
Marshes, 5
Melanism, industrial, 43
Mesozoic Era, 94, 99
Metamorphic rocks, 25
Millepora, 42
Mississippi river, sediments and, 8
Morphology:
  adaptive, 32
  functional, 32–36
Mud cracks, 15
Mutualism, in symbiosis, 49–50

N14 (*see* Nitrogen isotope)
Neptunists, 2
Neritic environment, 4
Newell, Norman D., 75, 107
Niches:
  biological, 31
  ecologic, 78–79
Nitrogen isotope, 89
North America:
  Cretaceous Period and, 82, 84,
    92–101
  glaciation, 84–91
  Ordovician Period and, 102–111
  Pleistocene Epoch and, 81–91
  Western Interior and, 92–101
North Atlantic Ocean Basin:
  chronology of, 81–91
  depth of, 82
  floor of, 82
  physiographic divisions, 82

Ocean (*see also* Cretaceous epicontinental sea; *specific examples,*
  e.g., North Atlantic)
  ancient seas, 64–66
  aphotic zone, 40
  carbon dioxide and, 39
  currents, 10–11
  environments of, 4
  North Atlantic, 81–91
  Ordovician seas, 102–104
  photic zone, 40
  salinity of, 40–41
  sediments, 10–11
  water depth, 72
  water turbulence, 41–43
Olson, E. C., 79–80
Omnivores, 49
Ordovician Period, 102–111
  British Isles, 106

Ordovician Period (*cont.*)
  Chazyan time of, 108, 110–111
  fossils, 106–108, 110–111
  reefs of, 107–108, 110–111
  seas during, 102–104
  sedimentary rocks of, 102, 106–
    108
  sediments of, 102–104
Organic compounds, ancient environments and, 66–67
Organisms:
  adaptive responses of, 30–32
  assemblages of, 53, 55
  carbon dioxide and, 38–39
  environments and, 30–60
  feeding types, 48–53
  functional morphology of, 32–36
  osmosis and, 41
  oxygen and, 38–39
  salinity and, 40–41, 55
  sediments and, 10–24
  sunlight and, 39–40
  temperature and, 36–38
  water turbulence and, 41–43
Orogeny, 92
  Taconic, 102
Osmosis, 41
Oxygen, organisms and, 38–39
Oxygen isotopes, 64–65, 87–88

Paleoecological analysis, 70–80
Paleoecology, 2–3
Paleontology, 3
Parasitism, in symbiosis, 49–50
Pelagic realm, of ocean, 4
Phenotype, 31
Photic zone, 40
Phytane, as "chemical fossil," 66–
  67
Pierre Shale, 96–97
Pitcher, Max, 107, 111
Playfair, Thomas, 70
Pleistocene Epoch, 81–91
  absolute chronology, 89–91
  glaciation during, 84–91
  North Atlantic Ocean during,
    81–91
  relative chronology, 84–89
Plutonists, 2
Principles of Geology (Lyell), 70
Pristane, as "chemical fossil," 66–
  67
Protactinium, isotope of, 91

Radiation, solar, 40
Reduction, problem of, 71–72
Reef barrier, 22
Reefs, 21–22
  coral, 73–75, 108
  as marine environment, 22
  Ordovician, 107–108, 110–111
Ripple marks, 13
Rivers, 5 (*see also specific ones,*
  e.g., Mississippi)
Rocks (*see also specific formations,*
  e.g., Pierre Shale)
  ancient marine, 65–66

GEOLOGIC TIME SCALE

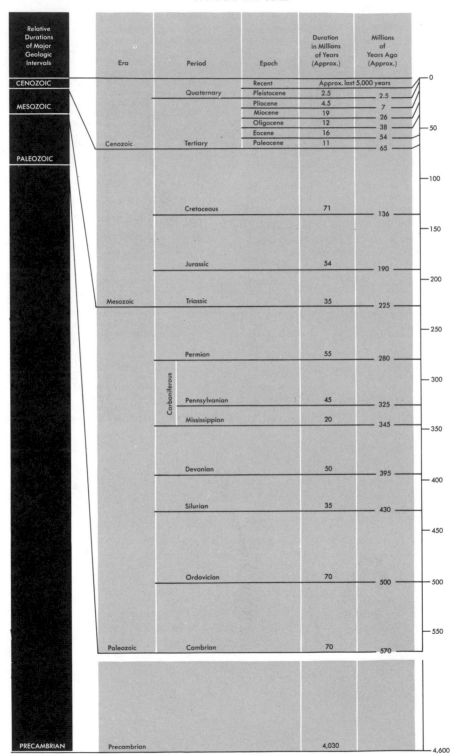

Relative Durations of Major Geologic Intervals	Era	Period	Epoch	Duration in Millions of Years (Approx.)	Millions of Years Ago (Approx.)
CENOZOIC	Cenozoic	Quaternary	Recent	Approx. last 5,000 years	0
			Pleistocene	2.5	2.5
MESOZOIC			Pliocene	4.5	7
			Miocene	19	26
			Oligocene	12	38
PALEOZOIC		Tertiary	Eocene	16	54
			Paleocene	11	65
	Mesozoic	Cretaceous		71	136
		Jurassic		54	190
		Triassic		35	225
	Paleozoic	Permian		55	280
		Pennsylvanian (Carboniferous)		45	325
		Mississippian (Carboniferous)		20	345
		Devonian		50	395
		Silurian		35	430
		Ordovician		70	500
		Cambrian		70	570
PRECAMBRIAN	Precambrian			4,030	4,600

Formation of Earth's crust about 4,600 million years ago

Millions of Years

Radiometric ages after Harland and others, 1964.